中国农业大学出版社“十三五”规划建设教材

能源电力英汉翻译案例讲评

皇甫伟　编著

中国农业大学出版社
·北京·

内 容 简 介

本书是翻译技能培养类实用书籍，可供能源电力专业教师和学生以及从事相关科技翻译研究的学者使用。本书也可作为教材，供学习者了解能源电力领域的专业知识内容、文本特点、翻译难点和常见对策，学习翻译实用技巧、翻译理论应用于科技翻译实践的方法。本书提供了丰富的译例和深入的译法分析，将理论与实践相互结合，对能源电力领域英汉翻译研究及应用具有较高的参考价值。

图书在版编目(CIP)数据

能源电力英汉翻译案例讲评 / 皇甫伟编著. —北京：中国农业大学出版社，2020.6

ISBN 978-7-5655-2363-2

Ⅰ.①能… Ⅱ.①皇… Ⅲ.①能源工业-英语-翻译②电力工业-英语-翻译 Ⅳ.①TK01 ②TM

中国版本图书馆 CIP 数据核字(2020)第 098958 号

书　　名 能源电力英汉翻译案例讲评
作　　者 皇甫伟　编著

策划编辑 张　玉　　**责任编辑** 张　玉
封面设计 郑　川
出版发行 中国农业大学出版社
社　　址 北京市海淀区圆明园西路 2 号　　**邮政编码** 100193
电　　话 发行部 010-62733489，1190　　读者服务部 010-62732336
编辑部 010-62732617，2618　　出　版　部 010-62733440
网　　址 http://www.caupress.cn　　**E-mail** cbsszs@cau.edu.cn
经　　销 新华书店
印　　刷 北京鑫丰华彩印有限公司
版　　次 2020 年 7 月第 1 版　　2020 年 7 月第 1 次印刷
规　　格 880×1 230　　32 开本　　6.75 印张　　190 千字
定　　价 59.00 元

本书献给我的妻子、儿子和我的父母

他们给予我前行的动力和无私的支持

前　言

规划“能源电力英汉翻译案例讲评”这个选题绝非我一时心血来潮，而是我的一个长期规划的一部分。如今本书作为出版社规划教材出版，可算是一个起点。我希望这部作品的完成能够成为一个良好的开端，自然不敢有半点马虎和敷衍。

我在华北电力大学讲授“科技英语翻译”已有多年，目前兴趣愈加浓厚，越讲越爱。我手头相关书籍和资料积累了不少，其中也不乏广为流传的名家名作和久经考验的实力之作。这些书吸引着我去阅读，帮我解惑，让我顿悟，启发我思考。当然，这些书皆为品质优良的佳作，这点我毫不质疑。同时，我也有这样的感觉，就是这些书比较适合喜好探索、享受学术阅读的外国语学院翻译类本硕学子或者翻译领域研究者使用。像我所在高校的“科技英语翻译”课程学习者多为学校非英语专业方向的研究生，让他们静下心来仔细阅读此类书籍，或者聆听使用此类书籍的教师的讲授，实在是强人所难。究其原因，原来这些书多采用语言学或者从英汉对比的角度探讨翻译技巧，编排的体例多有雷同。而且这些书的内容多偏重翻译技法，很少探究其内容的来龙去脉，使得学生知其然而不知其所以然。再有就是这些书多以词、短语和句子作为翻译单位讲解翻译技巧，译例的长度级别也鲜有达到句群或语篇的。尽管专门讨论科技、法律、经贸和政治等专门用途语言领域翻译的书也有不少，但多数所选译例并不集中在某个专业分支领域，这点尤其体现在科技英汉翻译类的书籍中。结果就是专业方向不同的读者碰到书中有异于自己专业类别的译例首先面临的就是专业语境的缺失，以及由此导致的理解和表达困难。如果是在平时的阅读中，读者完全可以通过使用专业工具或电子词典等手段解决问题。但如果是用在根本没有充足时间的教学场景，其使用效果就会打折扣。基于此考量，我觉得应该为读者提供包含更细分的专业领域、更长语段的翻译、更深入的原理分析、更丰富的背景知识介绍的实用翻译类图书实有必要，本书就是因此而作。

本书还有一个贯彻始终的主旨——译者能力的培养。这其中包括了译者的语言能力、专业知识能力、翻译能力、翻译理论应用能力和工具应用能力。英汉语基本功培养虽然不可能作为本书的重点内容，但是在每个章节的各部分内容里皆有不同程度的体现，比如怎么才能准确理解原文，如何进行信息结构整合和逻辑层次的表达等。专业知识能力的培养是本书的重点内容，包括了专业术语和专业背景知识两个方面，以让专业和非专业的读者群体都能够有所获益。翻译能力的培养是本书重中之重的内容，为此，本书采用了英汉语篇对照编排和翻译难点讲评的方式呈现翻译方法和技巧。翻译理论应用能力的培养是本书的特色内容，包括了对国内外不同理论倾向（原文和译文倾向）的多达十三种翻译理论在科技翻译实践中的应用解析。对译者工具应用能力的培养是本书紧跟时代发展步伐的体现，比如每个翻译难点和翻译理论应用的讲评或解析中，我都尽量向读者展示一些工具手段的使用，其中包括了词典工具、网络搜索工具、术语库和知识库等科技工具。

本书内容为专门用途英汉翻译案例讲评的能源电力类别，以后还可以此编排体例为标准，内容可再涉及其他专业科技学科领域，如自动化、计算机、农业、机械、建筑、生命科学等，还有人文社科的法律、经贸、政治等领域。本书中的译例都属于能源电力领域的相关内容，比如涉及有关风轮机、发电机、锅炉、电动机、核电、可再生能源、电力施工、电力设备、直流电源、不间断电源、锂离子电池等。每个章节分为原文、译文、术语表、翻译难点讲评、翻译理论及应用讲评和补充阅读六大部分，所以读者可采取部分选读或者全文阅读的方式，从每一部分汲取各自所需的有用信息。每篇原文长度约为一千字，提供翻译的语篇语境。英汉对照的编排有助于水平各异的读者自主选择阅读的方式，可以采用一边看原文一边看译文的即时对照方式，也可采用通读原文后再看译文的分时对照方式，也可两者结合进行，这比英语和汉语上下对照的编排更有利。本书按照原文出现的顺序为读者提供了术语表，方便读者查找关键术语。翻译难点讲评的撰写对我来说极具挑战性，也是最有意思的部分。在这里，读者可管中窥豹，找到我翻译译文的来龙去脉，其中文字多为我翻译过程中的心路历程记录。我在翻译理论及应用讲评部分，为读者精心选取了十三个翻译理论、翻译思想或翻译模式，其中有：信、达、雅，全译，生态翻译学，目的论，动态/功能对等理论，图式理论，翻译伦理，卡特福德翻译转换理论，翻译补偿理论，术语学，关联翻译理论，语义翻译理论和交际翻译理论，语境理论。对每种用到的理论，既有概要介绍理论的发展历程、理论的框架和要点内容，又通过文中具体的实例讲评其应用技法。在补充阅读里，为读者提供了更多的专业背景，并在本书的最后一章为每篇补充阅读统一提供了对应的译文，以方便读者自己翻译实战后对照参考。

本书是我在教学工作之余编写的，几乎占用了我全部业余时间，这对我来说极具挑战性，让我无法很好地担负起做儿子、丈夫、父亲的责任，应尽的家庭义务也没有完全尽到。我真心感谢家人的理解和支持。我还要诚挚感谢华北电力大学外国语学院的领导和同事对我工作的支持和指导，藉此才让我有机会完成本书的撰写。同时感谢中国农业大学出版社对本书出版给予的大力支持。

皇甫伟

于华北电力大学，北京

目　录

Chapter One Wind Turbine Transportation, Design and Installation

第一章 风轮机的运输、设计和安装

原文

1 Transportation and Design Impacts

The U. S. market has expanded to include lower wind speed sites (average wind speeds <7.5 m/s) closer to population centers. This is in part because of technological advancements and policy drivers. In some regions, it is also due to limited access to available transmission lines. As a result, from 1998 to 2013, turbines with larger blade and tower components can capture more wind at lower wind speed sites, but pose transportation and logistics challenges. The average estimated quality of the wind resource at 80 m for newly installed wind projects dropped by approximately 10%. This trend has increased the complexity and cost of transportation logistics because components such as blades and towers have increased in size to capture

译文

1 运输问题以及对设计的影响

美国风电市场已扩展到了靠近人口中心的低风速地区(平均风速小于7.5 m/s)。这一结果一部分是因为技术的进步与政策的推动,同时也是由于一些地区在并入输电线路上有一定局限。因此,在1998至2013年间,虽然具有较大叶片和风电塔结构的风轮机可以在低风速地区捕获更多的风能,但也带来了运输与物流上的挑战。为新风电项目供能的80米级风能的平均预估量下降了约10%。这一趋势增加了运输方面的复杂性和成本,因为,为了在较低风速地区捕获风能而加大了诸如叶片和风电塔之类的部件的尺寸。因此,现有的交通基础设施正日益影响着风电组件的设计,这些设计是为了实现能源发电与

the resource at lower wind sites. As a result, existing transportation infrastructure is increasingly impacting component designs to balance energy production with transportability.

可运输性之间的平衡。

1.1 Transportation Logistics

Installed turbine power ② ratings have continued to rise, to an average of 1.95 MW in 2012 including multiple models at more than 2 MW and above. As ③OEMs seek to capture more wind at lower wind speed sites, average rotor diameters have increased rapidly. Tower components have also increased in size and weight to access better winds higher above the ground. Wind turbine blades longer than 53 m begin to present a transportation obstacle due to the large turning radius, which hinders right of way or encroachment areas within corners or curves on roads or railways. Tower sections are generally limited to 4.3 m in diameter, or 4.6 m where routes permit, to fit under overhead obstructions.

In addition to the physical limitations associated with wind components, each state along a transportation route has different requirements to obtain permits.

⑤ This problem is exacerbated by

1.1 交通运输问题

现今使用的风轮机的额定功率在持续上升。至 2012 年,包括超过 2 兆瓦容量的多模型机组在内,风轮机的平均功率已达到 1.95 兆瓦。随着原设备制造商(OEM, Original Equipment Manufacturer)在低风速地区捕获更多的风能,风轮机转子的平均设计直径日益增大。风电塔组件在尺寸大小和重量上也有所增加,以获得地面以上更高处的更多的风能。长度超过 53 米的风轮机叶片由于大转向半径而开始对交通产生阻碍,比如会侵占到公路或者铁路上含有拐角或弯道的区域。风电塔横截面直径一般限制在 4.3～4.6 米,以适应道路区域与高架障碍物的限制。

除了与风电机组件相关的物理限制外,在运输线路上每一处通行许可的获得都有不同的要求。

由于货物体量的增大与风轮

higher volumes of shipments as wind turbine deployments increase. States are shifting the burden of proof for the safety of large, high-volume shipments to the wind industry. To address the increased complexity and resulting costs and delays associated with these logistics challenges, ③AWEA's Transportation and Logistics Working Group is coordinating with the American Association of State Highway and Transportation Officials to harmonize permitting processes across states. The increased size, mass, and quantity of wind components has resulted in more actively managed wind turbine transportation logistics, making use of a variety of land transportation methods and modes. ⑦This has resulted in increased project costs of up to 10% of capital costs for some projects.

机部署量的增加,这一问题更加严重。各个州正将对大体积货物运输发放安全许可的举证责任转移给风电行业。为了解决日益增长的复杂性问题、产生的费用问题以及这些运输相关的延误问题,美国风力能源协会(American Wind Energy Association,AWEA)的运输物流工作组正在与美国国家公路与运输协会进行协调,以协调跨州运输许可的问题。风力设备在体积、质量上的增加以及数量上的增长使风轮机在交通运输上的管理规划更为积极,这种运输利用了多种陆上交通方法和运输模式。这导致了一些项目成本的基本建设费用增加10%。

1.2 Design Impacts

Transportation constraints increasingly impact the design of wind turbine components, leading to higher capital costs resulting from sub optimal design. A prime example can be found in the industry-standard rolled steel wind turbine towers, which are limited to a structurally sub-optimal 4.3 m diameter to comply with size and weight limits of U.S. roads.

1.2 对组件设计的影响

运输上的限制越来越影响风轮机组件的设计,导致次优设计产生了更高的投资费用。在工业标准轧钢风轮机风电塔中可以找到一个最好的例子,这个风电塔为了符合美国道路的尺寸和重量限制,从结构上将次优直径限制在4.3米。虽然在这种直径约束下可以建造出高度为160米的风电塔,但

While it is possible to construct ① towers with hub heights up to 160 m at this constrained diameter, this height results in an exponential increase in the mass and cost of rolled steel towers. ⑩ Under transportation constraints as of 2014, tall towers are not economical in the sizes necessary to deploy wind in new, low and moderate wind speed land areas that are of interest to the industry to support cost reductions. It is important to note that these capital costs are substantially larger than the cost to transport the tower sections. Similar transportation-design trade-offs impact blades with respect to other aspects such as maximum chord dimensions. Some proposed solutions for on-site manufacturing of towers to mitigate transportation constraints can be found.

这个高度会导致轧钢风电塔的重量和成本呈指数增加。2014年，根据运输限制标准，使用大型风电塔在低风速和中风速地区部署风电并不经济，而在上述地区部署小型风电塔有助于成本降低，这对于风电行业具有吸引力。值得注意的是，这些投资成本比用在运输风电塔组件方面的成本大得多。在考虑到其他方面时，类似的运输设计权衡也会影响风机叶片的设计，如最大弦尺寸。现场建造风电塔可以减少运输方面的限制，可找到相关方案。

2 Installation

Because of the lift height and mass, hoisting a wind turbine nacelle onto its tower requires the largest crane capacity of all wind turbine construction and installation phases. The masses of a 3-MW nacelle assembly and a 5-MW nacelle assembly ⑨ are approximately 78 metric tons (t) and 130 t, respectively, without the gearbox and generator (104 t and 173 t with those components installed). Continued increases in tower heights and

2 安装

由于升降机的高度和质量，将风轮机吊舱吊装到风电塔上需要所有风力涡轮机建造和安装阶段的最大型起重能力。3-MW 机舱组件和 5-MW 机舱组件的质量分别约为 78 公吨(t)和 130 吨，而没有变速箱和发电机(安装了这些部件就变成了 104 吨和 173 吨)。风电塔高度和机器额定值的持续增加推动了机舱和桨叶质量的不断增加。因此，大型起重机的可用

machine ratings are driving higher nacelle and blade weights. As a result, the availability, scheduling, and logistics of larger cranes have become increasingly challenging. Alleviating this challenge could influence future wind deployment by facilitating cost-effective development in more regions of the United States. Analysis performed by ③ NREL indicates that having installation equipment capable of hoisting a 2.4-MW turbine onto a 140-m tower ⑧ would increase the economically deployable area for wind by 614,000 km^2 (237,000 mi^2), especially in the southeastern United States.

性、调度和流动性变得越来越具有挑战性。缓解这一挑战可能会影响美国更多地区的未来的风能部署,促进成本效益的发展。美国国家可再生能源实验室(National Renewable Energy Laboratory, NREL)进行的分析表明,安装设备能够将2.4-MW的风轮机吊装到140米的风电塔上,将使经济性可部署面积增加614,000平方千米(237,000平方英里),特别是在美国东南部。

④ Because mobile cranes capable of installing the majority of turbines deployed in the United States are of a common size used for construction and other industries, an ample supply of such cranes existed into 2014. As the number of turbines installed at 100 m hub heights and above has increased, however, concerns about the availability of larger capacity cranes has grown. There is the sharp drop in available U.S. cranes when shifting from the standard 600-ton to the 1,250～1,600-ton class cranes needed for taller towers and heavier nacelles.

在美国,能够安装大多数风轮机的移动起重机在2014年就已经普遍存在了,属于建筑业和其他工业所用起重机的普通型。由于安装在100米轮毂高度和以上的风轮机数量已经增加,然而,关于较大吨位起重机的可用性的担忧也已经增加。美国起重机从标准600吨到1,250～1,600吨级的可用数量急剧下降。大吨位起重机被用于更高的塔和更重的机舱。

⑥ Another challenge with larger crane classes is difficulty transporting

大型起重机带来的另一个挑战是将它们运输到风电场中并且

them to and maneuvering them within the wind plant, especially in complex terrain. A 1,600-ton crane has a width of nearly 13 m (41 feet), wider than a two-lane interstate highway (including shoulders), and requires more than 100 semi-tractor trailers to transport it between projects. This makes transportation between turbines difficult and costly.	操纵它们,尤其是在复杂地形中。一台 1,600 吨重的起重机,宽度约 13 米(41 英尺),比双车道州际公路(包括路肩)宽,并且需要超过 100 个半牵引拖车在项目工程之间来运输它,这就使得风轮机的运输困难,并且成本昂贵。

《术语表》

wind turbine	风力涡轮机;风轮机
blade	桨叶,叶片
logistics	后勤,物流
tower	风电塔
rolled steel	轧制钢材
nacelle	吊舱,发电机机舱
hub height	轮毂高度
mobile crane	移动式吊车;移动式起重机

翻译难点讲评

① 该翻译原文是能源类学术文献,涉及许多的风力发电的纯专业术语,例如①"tower, hub height"。这些专业术语有单义性和专业领域特有的特征,因此它们的翻译需要准确、专业化。译者在网络搜索查询,找到了风力发电领域的英汉对照的文本资料,《风力发电系统》是 2010 年由水利水电出版社出版的图书,译者是谢桦、王健强、姜久春,原作者是 Thomas Ackermann(德)(托马斯・阿克曼)。英文原文源自由 Thomas Ackermann 主编的 *Wind Power in Power Systems*,由 John Wiley & Sons, Ltd. 2005 年出版。从英汉对照文本中找到的上述专业术语分别译为"风电塔,轮毂高度"。

② 除了有许多的纯专业术语,原文也含有许多的半专业术语。半专业术语来自普通英语词汇,但在专业领域中使用时具有相应的专业词义,因此在

这些半术语的词义选择时必须考虑专业领域。比如，原文②中的"rating"，如果去查知网的网络翻译助手（dict. cnki. net），会得到"评级（513）、评定（982）、等级（248）"，括号内的数字是其使用的频率。这些译文都没有考虑"rating"的使用专业领域和上下文语境，应该是在能源电力领域，并且与"power"相关联的词义中进行选择。在百度翻译（fanyi. baidu. com）进行搜索，不但可以得到相关译文，而且还同时可以搜索到有该词出现的平行文本。可以发现，"rating"在能源电力中的译文多为"额定，定额"这两个，考虑到和"power"（功率）搭配，可顺序译为"功率定额"，也可逆序译为"额定功率"。

③ 有的情况下首字母缩略词的形式和意义联系不大，不懂其含义的人往往不知所措。有些科技方面的缩略词受到专业等的限制，仅仅在某个领域内使用广泛，当它们超出该领域进入大众视线时，普通人往往会有阅读和理解上的偏差。因此在译文中要规范这方面的首字母缩略词的使用，要注意：应该翻译出这些缩略词真正、准确的意思，并在第一次出现该词的译文时，如③中的"OEM""NREL"和"AWEA"，首先翻译出其意思"原设备制造商""美国国家可再生能源实验室"和"美国风力能源协会"，然后在括号内写出该缩略词本来的形式"Original Equipment Manufacturer""National Renewable Energy Laboratory"和"American Wind Energy Association"，并在这个短语后面注明其缩略形式。

④ 英文原文中出现使用显性连接词的句子，译文中连接词可以译出也可以省译。英文的衔接和连贯依赖于这些显性连接，而汉语是意合的语言，多使用隐形连接的方法达到连贯。即在表达各种逻辑关系时，英语句子内部的连接或句子间的连接采用句法手段（syntactic devices）或词汇手段（lexical devices），而汉语则多依靠语义的贯通，少用或不用连接词。例如句子④中的表达因果关系的"because"不作对应词翻译，在译文中没有翻译为"因为、由于"，让读者体会内在的逻辑关系，译文结构也因此更加简洁。

⑤ 被动语态广泛地出现在科技英语中，翻译时不可都采用"等值"顺译法或固守原句句式。在翻译科技英语被动语态语句时，要灵活地采用多种翻译技巧，必须根据汉语的语法和习惯，能译成主动句的就译成主动句，须译成被动语态的就译成被动句。这样既在内容上忠于原文，又符合汉语的表达方式。例如句子⑤中的"This problem is exacerbated…"如果依然

翻译为被动句"这个问题被……恶化",很难把"by"之后的内容都放在"被"字的后面。其实"by"和"as"后面的内容都是问题恶化的原因,不如都直接放在原因的位置,译成"由于货物体量的增大与风轮机部署量的增加,这一问题更加严重。"

⑥ 主位推进模式展现的是语篇中各小句的主位与述位如何相互关联、层层推进、进而汇句成篇,其体现了语篇中语言材料的排列组合形式及语篇中信息的流动与走向。分析原文的主位推进模式有助于译者准确地把握原作者的思路和解读原语篇,从而为构建通顺且忠实于原语语篇的译语语篇提供重要的参照。国内最早有朱永生(1995)主位同一型、述位同一型、延续型和交叉型的四分法比较侧重于主位推进模式理论在语篇分析中的应用。根据科技语篇的特点,有助于提高对科技语篇的理解和分析水平,可应用于科技文本的英汉翻译。于建平(2002)提出的五种主位—述位推进模式:主位简单线性(前一句的述位或述位的一部分成为后一句的主位)、主位同一推进(语篇中只有一个共同的主位,而每句的述位各不相同)、述位同一模式(主位不同,述位相同)、交叉推进模式(是前一句的主位是后一句的述位)、分裂述位推进模式(后面各句的主位由第一句的述位派生而来)。英汉两种语言在句子结构和表达方式上有各自的特点,译文不可能总是遵循原文的主位推进模式(李健、范祥涛,2008)。当保留原文的主位推进模式使得译文缺乏逻辑性和连贯性时,译者应在不损害原文信息结构的前提下,重建译文的主位推进模式,使之读起来自然顺畅,达到原文整体的语篇效果,其中主要采用语序调整法和重组法。段落⑥中是两种推进模式的组合,第 2 和第 3 句子之间是主位简单线性推进,而第 1 句和 2 句之间则是主位统一推进模式。译文中,首先把 1 和 2 句进行了合译重组,并把第 1 个句子中的主述位位置进行了调整,从而使其推进模式转变为主位同一型,这种转变恰当地构建了符合汉语语言习惯的主位推进模式。

翻译理论及应用讲评:信、达、雅

"信、达、雅"(faithfulness, expressiveness, elegance)是严复在赫胥黎《天演论》中的译例言中首次提出的翻译思想,也可作为其科技翻译实践的经验总结。后来学者依据时代发展对严复的思想进行研究,确立了信、达、雅严复翻译理论的地位(沈苏儒,1998)。严复对翻译标准的真知灼见,“信、达、雅”三字真言大体上也就相当于今天科技翻译的通用标准的“准确、通畅、简洁”六字标准。

张树柏在《谈科技论文的翻译》中提出了三个翻译要求:本意保留、风格一致、语言规范,实质上与严复不谋而合。张忠华在《科技英语汉译的“达”和“雅”》中提出科技英译汉的第一质量标准是准确,即“信”,而“流畅通顺、简洁明快”,即“达”和“雅”是科技译文质量的又一个方面。罗新璋《我国自成体系的翻译理论》也承认了“信、达、雅”对科技翻译的指导作用,分别体现在忠于原文、文体对等、译笔优美。也有马谷城用“易懂”代替“雅”。但是,“译笔优美”能否在用词客观、行文简洁、逻辑紧凑,并且常常有复杂句式的科技英译汉中实现,这是一个难题;“易懂”也忽视了科技英语文本自身的语言特点。严复的翻译思想也与爱丁堡大学的泰特勒(Tytler)提出的翻译原则高度契合:“That the translation should give a complete transcript of the ideas of the original work; that the style and manner of writing should be of the same character with that of the original; that the translation should have all the ease of the original composition.”下面结合本章文本和译文举例说明。

1. 信/faithfulness

例⑦“This has resulted in increased project costs of up to 10% of capital costs for some projects.”译文为“这导致了一些项目成本的基本建设费用增加达 10%”。此处的表数量增长的“increase...up to...”不能译为“增加达到”。正文文本另外一个例子⑧也是同样道理:“...would increase the economically deployable area for wind by...”。这里也只能译为“将使经济性可布置面积增加……”而不是“……增加到……”。正文中还有一处力求忠于原文的形式与内容的例⑨是“...approximately 78 metric tons (t) and 130 t respectively...”。译为“……分别约为 78 公吨(t)和 130 吨……”。因为 1 公吨(metric ton)= 1 吨(t)。中国采用公制单位,所以“吨”指的就是“公吨”,但是译文依然从“信”的基本点出发,对原文不同的表达作出恰当的译文选择。

2. 达/expressiveness

科技英语中有大量的技术词汇、半技术词汇和缩略语,这符合科技英语专业性和学术性的文体特征。这些词语的翻译必须意思表达准确、规范、地道,不可胡译乱译为文体不对等、表意不清晰的译文,让专业人士摸不着头脑。比如，在本章文本中"logistics"是"物流"，而不是"逻辑"；"tower"在文中是"风电塔"，而不能翻译为"塔"；"nacelle"是"发电机机舱"而不是"吊篮"。另外，AWEA是"American Wind Energy Association"的缩写，译为"美国风力能源协会"，NREL是"National Renewable Energy Laboratory"的缩写，译为"美国可再生能源实验室"。这些术语和缩写的使用是科技英语文章扩大信息量、简化表达的手段，其译文必须符合专业化的要求和语言表达的规范。

3. 雅/elegance

科技英译汉中的"雅"集中体现为译文要符合原文的行文特征，即逻辑紧凑，行文简洁，避免语焉不详、单调和重复。比如例⑩"Under transportation constraints as of 2014, tall towers are not economical in the sizes necessary to deploy wind in new, low and moderate wind speed land areas that are of interest to the industry to support cost reductions."译文为"2014年，根据运输限制标准，使用大型风电塔在低风速和中风速地区部署风电并不经济，而在上述地区部署小型风电塔有助于成本降低，这对于风电行业具有吸引力。"此句主干是"... tall towers are not economical in the sizes..."，之前有介词短语作状语，之后有复杂的后置定语"... to deploy..."和"... in..."，以及后面的that从句。要使译文"雅"必须把逻辑意思表达清晰，即风电塔的大小与中低风速地区新部署的风力发电的经济性，以及风电塔的大小对于风电行业利害相关的是进一步降低投资成本。所以一切围绕"风电塔大小"展开，并且"... sizes ... that..."结构里先行词与定语从句被割裂开，译文在从句处分译出另外一个单独的小句，让译文中使用短句，避免了使用长句容易造成的语义不明，也更加符合汉语的表达规范。

补充阅读:Steam Turbine

The steam turbine shall be a single-shaft, tandem compound, 3000 RPM, single-reheat condensing turbine suitable for indoor installation. The steam cycle arrangement for the Power Station, including turbine generator parameters, shall be selected by the Contractor to achieve the Guaranteed Net Electrical Output at the high side of the GSU at rated main steam flow, pressure, and temperature conditions.

The steam turbine shall have an additional 3 percent (3%) (minimum) flow margin above this point with valves wide open (VWO). The turbine generator and the rest of the plant shall be capable of continuous long term operation at VWO, normal pressure conditions. The cycle conditions, including the number of feed water heaters, approaches, terminal temperature differences, heater drain arrangement, backpressure, last stage blade selection, and other steam cycle parameters shall be optimized by the Contractor based on rates as measures of the relative value of capital cost and operating cost.

The steam turbine shall be a tandem compound reheat machine consisting of a high-pressure turbine (HP turbine) and intermediate-pressure turbine (IP turbine) casing and four or six flow low-pressure turbine (LP turbine) casings exhausting downward into the surface condenser.

The turbine casings shall be horizontally split along the turbine rotor centerline.

The turbine rotors shall be statically and dynamically balanced. A complete system of valving to protect the turbine from overspeeding shall be furnished.

The design features to reduce and control thermal stresses in the turbine shall include:

1. Full arc admission capability at startup using control valve configuration only, not the turbine stop valves.

2. Sliding throttle pressure operation from minimum load to 100 % MCR-pressure condition and then fix pressure to VWO (103 % MCR) condition.

3. A rotor stress monitoring system to assist in startup by monitoring and presenting rotor stresses in terms of cycle life expenditure during steam-to-metal temperature matching, acceleration, and loading.

The turbine blading and the turbine generator control system shall be designed to allow for low-frequency and high-frequency operation commensurate with the Grid Code requirements listed in Attachment A. Startup and initial loading of the turbine shall be controlled by the turbine control and instrumentation system by either automatic or fully manual means with operator guidance that follows the Original Equipment Manufacturer's "OEM" recommended procedures.

All protective tripping functions shall be implemented using redundant sensors, so that the failure of one device does not make the turbine system inoperable.

Self-diagnostics shall be provided to identify devices that are out of normal operating range. The self-diagnostic system shall alarm the out-of-range devices so that they can be identified for repair.

Functional constraints applicable to the steam turbine generator shall be identified by the turbine manufacturer for such conditions as startup and loading procedure, speed, condenser pressure, throttle and reheat temperature and pressure, and steam quality. The turbine manufacturer shall provide startup and operation curves for starting the turbine in the cold, warm, and hot conditions.

The turbine lube oil system shall include, as a minimum: one shaft driven main oil pump and one AC oil pump, DC emergency oil pump, two full-capacity duplex lube oil coolers, one vapor extractor mounted on the reservoir, an oil reservoir (clean lube oil tank), a lube oil conditioner, and a dirty lube oil tank.

One main lube oil pump shall be turbine shaft driven, with a back-up full flow AC motor driven pump. All other pumps required for system operation shall also be provided. All motor driven lube oil pumps shall be mounted on the lube oil reservoir. Each oil pump suction and oil return line shall be provided with a strainer.

There shall be no emergency stop push button on the emergency oil pump for the turbine.

All lube oil piping shall be 18-8 stainless steel and guarded. The guard pipe may be carbon steel. Lube oil piping shall be routed in a manner to avoid crossing or being located directly above any steam piping, sloped to avoid air pockets and allow gravity drainage where possible, and within the confines of the turbine foundation where possible. Redundant pressure transmitters shall be provided for automatic sequential starting of the motor driven oil pumps. Individual pump discharge pressure transmitters shall be provided to indicate operation of each individual oil pump. Provisions shall be included for testing of the AC bearing oil pump(s) and DC emergency oil pump auto-start from the control room. Pressure transmitters shall also be provided for the common lube oil supply header and the individual lube oil supply to each bearing.

A turbine lube oil purification system, including a circulating pump, pressure particulate filter, and coalescing water removal filter shall be provided. This system shall remove particulate matter down to 10 microns or less and shall remove any free water in the oil being processed. The system shall be designed to treat at least 20-percent of the lube oil system capacity per hour.

The turbine lube oil system shall also be furnished with a clean and a dirty lube oil tanks, each with a working capacity equal to the total turbine lube oil reservoir capacity and a level transmitter for control room level indication.

The turbine exhaust hood water spray system shall be a complete system and shall be the supplier's standard for protection against high exhaust hood temperature.

The gland seal system and gland condenser shall be the turbine supplier's standard. Seal steam shall be provided from the Auxiliary Steam System, with bypass valves provided around the seal steam feed valve and unloading valve.

Chapter Two The 2030 Goal of Universal Energy Access

第二章 普及能源接入2030年目标

原文

Despite the serious efforts already made, today an estimated 1.2 billion people - 17% of the global population—remain without electricity, and 2.7 billion people - 38% of the global population—put their health at risk through reliance on the traditional use of solid biomass for cooking. ⑧The newly agreed UN Sustainable Development Goals embrace a goal on energy, a move long advocated by the IEA, including the target to achieve universal access to energy by 2030. In our outlook, the number of people without electricity falls to 800 million by 2030 and the number without access to clean cooking fuels declines only gradually to 2.3 billion in 2030.

译文

尽管已经做出了很大的努力,至今估计有12亿人(占全球人口的17%)仍然没有用上电,27亿人(占世界人口的38%)通过依靠传统的固体生物质燃料来做饭,这使他们的健康处于危险之中。新达成的包括能源目标在内的“联合国可持续发展目标”是国际能源机构(International Energy Agency ,IEA)长期倡导的目标,其中包括到2030年前实现普遍获取能源的目标。照我们的预期,到2030年用不上电的人口将减少到8亿人,没有清洁燃料做饭的人口将会逐渐下降到23亿人。

1 Risks Along With Higher Oil Prices

1 高油价背后的风险

Oil prices head higher as markets work off the excess supply, but risks

虽然市场摆脱了供应过量,石油价格上涨,但是风险仍然存

remain: ③ The process of adjustment in the oil market is rarely a smooth ③ one, but, in our central scenario, the market rebalances at $80/bbl in 2020, with further increases in price thereafter. Demand picks up to 2020, adding an average of 900 kb/d per year, but the subsequent ④ rise to 103.5 mb/d in 2040 is moderated by higher prices, efforts to phase out subsidies (provided that momentum behind reform is maintained, even as oil prices pick up), efficiency policies and switching to alternative fuels. Collectively, the United States, EU and Japan see their oil demand drop by around 10 mb/d by 2040. On the supply side, the decline in current upstream spending, estimated at more than 20% in 2015, results in the combined production of non-OPEC producers peaking before 2020 at just above 55 mb/d. ⑤ Output growth among OPEC countries is led by Iraq and Iran, but both countries face major challenges: the risk of instability in Iraq, alongside weaknesses in infrastructure and institutions; and the need in Iran (assuming the path to sanctions relief is followed successfully) to secure the technology and large-scale investment required. An annual $630 billion in worldwide upstream oil and gas investment—the total amount the industry spent on average each year for the past five

在，石油市场的调整过程很少是一帆风顺的，我们的中心愿景是在2020年市场以每桶80美元的价格重新平衡，然后价格会进一步上涨。至2020年石油需求持续回升，年均增加90万桶，但是，在价格的上涨，以及逐步取消补贴的作用（只要将改革的势头保持下去，即使油价回升）和采取能效政策并转向替代燃料等措施调控下，随后上升到2040年的1035万桶/日是有可能的。总的来说，美国、欧盟和日本的石油需求到2040年预计下降大约1000万桶/日。在供应方面，目前上游支出的下降（2015年估计超过20%），这将导致非欧佩克产油国的产量总和在2020年以前达到略高于55万桶/日的峰值。伊拉克和伊朗主导欧佩克国家的产出增长，但这两个国家都面临着重大挑战：伊拉克存在不稳定性以及基础设施薄弱和政策无力的风险；而伊朗（假如解除制裁的路线图成功实施）需要确保技术得到满足和大规模投资到位。全球每年有6300亿美元用于石油上游业务和天然气投资——这是该行业在过去5年里平均每年花费的总额——只是为了弥补油田产量下降的损失并保持未来的产

years is required just to compensate for declining production at existing fields and to keep future output flat at today's levels. The current overhang in supply should give no cause for complacency about oil market security.

量在今天的水平。当前的供应过剩不应该让人们对石油市场的安全感到自满。

The short investment cycle of tight oil and its ability to respond quickly to price signals is changing the way that the oil market operates, but the intensity with which the tight oil resource is developed in the United States eventually pushes up costs. US tight oil production stumbles in the short term but resumes its upward march as prices recover, helped by continued improvements in technology and efficiency improvements. ⑦ But tight oil's rise is ultimately constrained by the rising costs of production, as operators deplete the "sweet spots" and move to less productive acreage. US tight oil output reaches a plateau in the early-2020s, just above 5 mb/d, before starting a gradual decline.

致密油短期投资周期及其对价格信号迅速反应的能力正在改变石油市场运作的方式，但是致密油在美国的发展强度最终推高了成本，美国的致密油产量在短期内出现了下滑，但随着价格回升，以及技术和效率的持续进步和改善，产量恢复了上行趋势。但是，因为运营商耗尽了"甜点"转向生产效率低的区域，生产成本上升，致密油的崛起最终受到了限制。美国致密油产量在21世纪20年代初期达到了略高于500万桶/日的一个稳定的水平，然后开始逐渐下降。

What if prices stay lower for longer? A more prolonged period of lower oil prices cannot be ruled out. We examine in a Low Oil Price Scenario what it would take for this to happen—and what it would mean for the entire energy sector if it did. The oil price in this scenario remains close to $50/bbl until the end of this decade, before rising gradually back to $85/bbl in

如果价格持续走低，会怎么样呢？我们不能排除更长期的低油价的情况。我们研究在一个低油价的情况下会有什么样的后果——如果油价持续走低，这对能源部门意味着什么。在这一情况下，直到2020年石油价格仍将接近50美元/桶，然后在2040年逐步回升至85美元/

2040. This trajectory is based on assumptions of lower near-term growth in the global economy; a more stable Middle East and a lasting switch in OPEC production strategy in favor of securing a higher share of the oil market (as well as a price that defends the position of oil in the global energy mix); and more resilient non-OPEC supply, notably from US tight oil. With higher demand, led by the transport sector, pushing oil use up to 107 mb/d in 2040, the durability of this scenario depends on the ability and willingness of the large low-cost resource-holders to produce at much higher levels than in our central scenario. In the Low Oil Price Scenario, the Middle East's share in the oil market ends up higher than at any time in the last forty years.

桶。这一轨迹是基于如下假设：全球经济短期增长放缓、一个更加稳定的中东和欧佩克生产战略的持续转变有利于获得更高的石油市场份额(以及维护全球能源结构中石油地位的价格)；更有弹性的非欧佩克供应，尤其是美国致密油的假设。在运输部门的带动下，需求会增加，到2040年石油消耗将达到10700桶/日，这一情况的耐久性取决于大型低成本资源持有者是否有能力和意愿提高产量，使他们的生产能力远高于我们的中心愿景。在低油价的情况下，中东在石油市场的份额最终比过去40年任何时候都要高。

2 Turbulent Times Ahead For Coal

Coal has increased its share of the global energy mix from 23% in 2000 to 29% today, but the momentum behind coal's surge is ebbing away—and the fuel faces a reversal of fortune. Expectations within the industry of continued strong demand growth, triggered major investments in supply in recent years but actual coal use has fallen well short, resulting in over-capacity and plummeting prices. In our projections, the fuel that met 45% of the increase in global energy demand over

2 煤炭市场前景动荡

煤炭在全球能源结构中的份额已从2000年的23%增加到现在的29%，但煤炭飙升的势头正在消退，煤炭的命运甚至将面临逆转。煤炭行业对需求持续强劲增长的预期近几年引发了大量的供应投资，但实际煤炭消耗量却大幅下降，导致了煤炭产能过剩和煤炭价格暴跌。根据我们的预测，满足过去十年全球能源需求增长45%的煤炭仅满足目前至2040年的进一步增

the last decade meets only around 10% of further growth to 2040, largely due to a tripling in coal demand in India and in Southeast Asia. Consumption in the OECD, where coal use faces ①strong policy headwinds, is projected to drop by 40% over the same period: coal consumption in the European Union in 2040 falls to around one-third of current levels. From a position as a perceived ② safe bet, China is becoming the ②wild card of coal markets, with the risks to our projection of a plateau and then a slow decline in coal demand arguably weighted to the downside. ⑥By 2040, Asia is projected to account for four out of every five tons of coal consumed globally, and coal remains the backbone of the power system in many countries in our central scenario. However, its continued use around the world is compatible with stringent environmental policies only if it is used in the most efficient way, with advanced control technologies to reduce air pollution, and if progress is made in demonstrating that CO_2 can be safely and cost-effectively captured and stored.

长需求的10%左右,这主要是由于印度和东南亚的煤炭需求增加了三倍。在经济合作与发展组织(Organization for Economic Co-operation and Development, OECD),煤炭消费面临强大的政策顶头风,消费预计将同比下降40%。2040年欧盟的煤炭消费量下降至目前水平的1/3左右。中国从一个稳定的煤炭消费国,正在成为煤炭市场的变量,我们对该国煤炭需求保持高位平稳趋势,随后缓慢下降,实际可能处于下行。到2040年,亚洲煤炭消耗量预计将占全球的4/5,而煤炭仍是我们中心愿景中许多国家电力系统的支柱。但是只有通过采用先进的控制技术减少空气污染,来使煤炭获得最高效的利用,实现CO_2安全且节约成本地捕获及存储,煤炭在全球范围内的持续使用才能满足严格的环境政策要求。

3 The Power Sector is Leading the Charge to Decarbonize

3 电力部门正在引领脱碳

Electricity gains ground in many end-use sectors, making up almost a quarter of final energy consumption by 2040; the

到2040年,电力将在许多终端能源用户部门得到增长,几乎占到最终能源消费的1/4;电

power sector leads the way towards a decarbonized energy system. Non-OECD countries account for seven out of every eight additional units of electricity demand. ⑨ With 60 cents of every dollar invested in new power plants to 2040 spent on renewable energy technologies, global renewable-based electricity generation increases by some 8 300 TWh (more than half of the increase in total generation), equivalent to the output of all of today's fossil-fuel generation plants in China, the United States and the European Union combined. The net result is that the share of coal in the global electricity mix drops from 41% to 30%, with non-hydro renewable gaining a similar amount, while gas, nuclear and hydro broadly maintain their existing shares. By 2040, renewable-based generation reaches a share of 50% in the European Union, around 30% in China and Japan, and above 25% in the United States and India; by contrast, coal accounts for less than 15% of electricity supply outside of Asia. Despite both more costly technologies and rising fossil-fuel prices, electricity is set to become more affordable, relative to GDP, in most regions. With more generation from renewable and nuclear power, and more efficient thermal plants, CO_2 emissions from power generation are set to grow at

力部门将引领走向脱碳能源体系的道路。非经济合作与发展组织国家占额外电力需求单位的7/8。到2040年，当每1美元中就会有60美分用于新电厂可再生能源技术的投资，全球可再生能源发电量会增加约8300太瓦时，占总发电量增长的一半以上，这与当今的中国、美国和欧盟的化石燃料发电厂加在一起的总发电量相当。最终的结果是煤炭在全球电力结构中的份额从41%下降到30%，非水力可再生能源增加相似份额，而天然气、核电和水电大体维持现有份额。到2040年，可再生能源发电在欧盟将达到50%的份额，在中国和日本占30%左右，在美国和印度占25%以上：相比之下，在亚洲之外，煤炭占电力供应的份额不到15%。尽管技术成本更高且化石燃料价格上涨，但是相比较于多数地区的国内生产总值，电力的消费变得更能承担得起。随着可再生能源和核能发电量的增加以及更高效的热电厂的出现，到2040年电厂的二氧化碳排放量增速将只是电力发电量增速的1/5；而过去25年来两者是1:1。为了实现这些预期，世界需要在年前增加比现在全球更多的装机

only one-fifth of the rate at which power output rises to 2040; this was a one-to-one relationship over the last 25 years. To realise these projections, the world needs to add more capacity by 2040 than is globally installed today, while average utilization rates for capacity go down because of the need to integrate variable renewable energy technologies. This raises questions in many countries about the appropriate market mechanisms that can generate the necessary investment in generation and grids.

容量,而由于需要整合波动性可再生能源技术,因此平均装机容量利用率下降。这对许多国家提出了关于建立适当的市场机制的问题,以为发电和电网提供必要的投资资金。

《术语表》

energy access	能源接入
solid biomass	固体生物质燃料
kb/d; mb/d	千桶/日;百万桶/日
tight oil	致密油
decarbonize	脱碳
TWh.	太瓦时(10亿千瓦时)
fossil fuel	化石燃料
renewable	可再生能源
thermal plant	热电厂
capacity	装机容量

翻译难点讲评

① 该句中,“strong policy headwinds”也可以翻译为“强大的政策阻力”。这里按照词语本身的含义直接译成“强大的政策顶头风”,而不是将其内涵意思译出来,这样译可以更加生动形象地将煤炭所遭遇的状况表达出来,便于读者形成直观的印象,方便读者理解,既保持原文内容、又保持原文形式。而且,“顶头风”可能对于外国人其含义多为其字面意思“逆风、迎着

风”，而对于中国人来说其还含有隐喻意义“比喻不顾恶劣的境遇”。况且，这里是煤炭消费市场迎来了政策的寒冬，“漏船偏遇顶头风”，说其顶风、负重，艰难前行，也是非常形象的。

② 句子中“bet”的意思是赌注，很显然将“safe bet”按词义本身翻译“准能赢的打赌”并不正确。对照后面的“the wild card”本义“计算机中的通配符，可用于替代任一字符或系列字符一词”，或者纸牌游戏中的“百搭牌，由持牌人决定牌值的牌”。所以结合上下文语境，“a plateau and then a slow decline in coal demand arguably weighted to the downside”，说“煤炭需从高位平稳，随后缓慢下降，实际可能处于下行”，据此来看，中国是从煤炭市场中的一个稳定的煤炭消费国变成了一个变量。这样翻译既忠实于原文，前后对照，合乎语境又符合汉语的表达方式和思维习惯，翻译得更加通顺、准确，增强了文章的可读性。

③ 句中“the process of adjustment”按照字面意思可翻译成“调整的过程”，“one”是个代词，分析句子可知其指代的也是“the process of adjustment”。如果逐字翻译那么就是“石油市场的调整过程一般都不是一个平稳的过程”这样翻译显然非常啰唆，也不符合汉语表达习惯。需要删去不符合目标语思维习惯、语言习惯和表达方式的词，以避免译文累赘。所以根据语境分析并按汉语的理解，根据汉语的表达习惯，采用了省词译法，将原文的“one”省去了翻译。这样翻译会使译文更加简洁，更符合汉语的逻辑，继而达到简洁流畅的目的。

④ 文中“rise”是名词，采用转译法，把原文的名词译成译文的动词，翻译成“增长”。英语是静态语言，多用名词，汉语是动态语言，多用动词（连淑能，2010）。由于英汉两种语言的差异，将名字译作动词更能动态地表达原文所强调的石油消费量增长的意图和目的，凸显文意，符合汉语的表达习惯，使译文语言特点鲜明，准确且形象地表达了原文的意思，连贯通顺。汉译时要注意英语和汉语动词的特点，往往可以将英语中的名词、前置词、形容词和副词转译成汉语的动词。不但忠实于原文意思，还贴合了汉语的思维方式，更能够在忠实于原作的基础上，准确、连贯并且流畅地将意思传达出来，便于读者理解。

⑤ 原句中“is led by”，“the path is followed”，“investment required”是被动或者逻辑被动形式，在英语中很常见，但是如果直接采用被动形式进行翻译的话，就是“欧佩克国家产量增加是由伊朗和伊拉克主导”，这样翻译有生

拉硬拽之嫌，而且也并不符合汉语的表达习惯。在英语的被动句里，如果没有逻辑主语，或者虽有逻辑主语，但要强调它在译句里充当主语或宾语的作用，又或者汉语不能或不宜用被动句表示，那就要把这被动句译成主动句。在汉语中一般会采用主动的形式来表达被动的含义，所以根据汉语多主动的表达习惯。采用语态转换译法，对其进行翻译，将“is led by 由…主导”转译为主动的“主导……”，“the path is successfully followed”转译为“路线图成功实施”，“investment required”转译为“需要投资”。译文更加地道自然，连贯性好。

⑥ 原文由三个句子构成，第一个是“Asia... globally”，第二个是“coal remains... scenario”，第三个是“However, its... pollution”。对句子进行分析，第一句的“coal consumed globally”是宾语，第二句的主语是“coal”，第三句的主语是“its continued use”，而其中的 its 指的是 coal's。考虑到三个句子中都提到了“coal”（煤炭），所以为了营造主位相同而述位各不相同的主位同一性推进模式，使得译文句式符合汉语语言连贯的原则，采用了合译法将第二句和第三句合成一个句子进行翻译，同时把第一句的宾语迁移到主语的位置。这样翻译使译文的在衔接上更加紧密，在内容上紧凑充实，读起来朗朗上口，简洁通顺地表达了原文意思。

⑦ 该句的主干是“tight oil's rise is constrained”“by the rising costs of production”是介词短语用来修饰动词，“as operators deplete”是一个从句。英语是形合语，汉语是意合语，英译汉时通常先将修饰成分和从句等优先译出来，然后再翻译主干部分。其实本句中的介词短语和从句都是主句所说“致密油价格上升的原因”，所以按照汉语的因果关系，先因后果，采用换序译法，先翻译从句，然后再翻译其他部分。不仅忠实地表达了原文意思，而且翻译起来也更加灵活多变。这样译文更加通顺流畅，避免了翻译的机械性，也更符合汉语的表达习惯。所以把“But tight oil's rise is ultimately constrained by the rising costs of production, as operators deplete the ‘sweet spots’ and move to less productive acreage.”翻译为“但是，因为运营商耗尽了‘甜点’转向生产效率低的区域，生产成本上升，致密油的崛起最终受到了限制。”

翻译理论及应用讲评：全译

全译（complete translation）作为一个翻译范畴的提出，与变译相对，由黄忠廉在他的《变译理论》一书中提出，后他的学生余承法在名为《全译求化机制论》的博士论文中将其发扬光大。全译的本质是求化，即力求与原文语表形式、语里意义和语用价值高度契合或一致。全译求“化”追求的是译文和原文在意似和形似的基础上的风格相似，遵守语用价值为核心，语里意义为基础和前提，语表形式为结果和外化的原则。由此，全译求“化”即为传原文之意，存原文之味，成译文之形。与变译求“变”不同，全译求“化”的运用必须把首要目标设定为再现原文语用和准确表达原文语义，可以通过改变或者保留原文语形来实现。具体方法可分为等化、繁简(繁化＋简化)、移换(移化＋换化)、分合(分化＋合化)四大类，共七种求化机制。每类求化机制还可向下细分，根据操作的语言单位可分为30个具体的方法：词/短语/小句/复句等化；语素/词/短语/小句繁化；词/短语/小句/复句简化；原形/有形/无形移化；词类/成分/句类/动静/正反/主被转化；词/短语/小句/复句分化；短语/小句/复句/句群合化。除等化外，其他全译方法都有一定程度的“变”，但全译之“变”大都在句内完成，很少涉及句群以上的单位“语篇”。如有涉及也只是关乎衔接和连贯手段，如指代和照应等。以上全译方法可单用，也可合用。下面从本章文本中选取例子对部分类型的全译方法加以说明。

例⑧“The newly agreed UN Sustainable Development Goals embrace a goal on energy, a move long advocated by the IEA, including the target to achieve universal access to energy by 2030.”新达成的包括能源目标在内的“联合国可持续发展目标”是国际能源机构(International Energy Agency, IEA)长期倡导的目标，包括到2030年前实现普遍获取能源的目标。此处原文主干是“...Goals embrace a goal...”，逗号后面的“a move...”是“a goal”的同位语，最后的“including”短语可以看作是介词短语或者分词短语。译文主要采用了有形移化和成分换化的机制，即把主干的谓语和宾语转换为主语的定语的一部分，把同位语转换为系表结构成为复合谓语。当然换化的同时，原文的部分语形也发生了空间位置的移动，例如“embrace a goal on energy”就移动到了“UN...”之前，这在译文中也是为了遵守汉语的对名词前定语的位置的表达习惯和规范要求。此例也显示出移化是一种单向行为，换化则为双向，两者可共同应用于同一翻译单位，既转换其表达方式，也改变其位置，两者相互补充，共同语义求真、语用求美。

例⑨“With 60 cents of every dollar invested in new power plants to 2040 spent on renewable energy technologies, global renewable-based electricity generation increases by some 8 300 TWh (more than half of the increase in total generation), equivalent to the output of all of today's fossil-fuel generation plants in China, the United States and the European Union combined."译文为"到 2040 年,当每 1 美元中就会有 60 美分用于新电厂可再生能源技术的投资,全球可再生能源发电量会增加约 8300 太瓦时,占总发电量增长的一半以上,这与当今的中国、美国和欧盟的化石燃料发电厂加在一起的总发电量相当。"此例主要采用了繁化和分化的机制。增补了原文语义含有的,语形没有的语言单位。包括增补了表示时间的"当……",增补了限定副词"……就……",对应翻译原文中的"with..."结构。增补了原文没有的表示将来的"会……",以及把括号内的内容通过增加谓语动词"占",把"equivalent..."补足语中的内容通过增加主语"这",都分化为一个与主句并列的单句。此例显示出繁化机制是一种译文显化的过程,即语言单位从原文到译文的从无到有,从少到多,从短到长。这与简化机制的操作方式相反,简化是一种隐化的过程,即语言单位从原文到译文的从有到无,从多到少,从长到短。繁化常常伴随分化,因为在拆分原文语形后常常要补足一些成分才能保证原文语用价值再现,语义传达和译文表达之需。

补充阅读:Boiler

The system design must meet all the requirements of the Plant. The boiler shall be of advanced design and shall be based on supercritical technology with sliding pressure operation, once-through type, single reheat, single furnace, two fireballs, equipped with economizer and regenerative air heaters, vertical tube arrangement for the furnace wall, balance draft, pendant structure steel frame construction.

The boiler shall be an outdoor, pulverized coal fired, balanced draft, once-through, unit equipped with an economizer and regenerative air heaters. The boiler shall be designed, constructed and performance tested in accordance with, amongst others, ASME Boiler and Pressure Vessel Code, ASME B31.1, NFPA and Factories and Machinery Act 1967 (Act 139) and Regulations.

To control the effect of slagging and fouling, the Contractor shall provide a complete soot blowing system for the various sections of the boiler and detailed drawings showing the portion of the equipment, angle of blow and any spare holes on the furnace walls.

The boiler structure shall be provided with an uninsulated roof and siding extending down ten (10) m as measured from the boiler topmost structure. The boiler shall be one (1) × 100 percent (100%) capacity unit designed to burn any coals listed in coal specifications of the Employer's Requirements, either alone or in combination, using the manufacturer's standard low NO_x burners and over-fire air arrangement. Combustion air shall be provided by primary air fans and forced draft fans. The boiler shall be top supported and shall be a two-pass design to be determined by the Contractor.

The boiler and auxiliaries shall be designed for safe and reliable operation from startup to the boiler maximum continuous rating (BMCR) of the unit. The unit shall be capable of stable firing without the use of supplemental fuel, from 30 percent (30%) load to BMCR. The startup fuel shall be distillate fuel. The Power Station shall be designed for continuous safe operation in a controlled, stable manner to facilitate plant operation from minimum load without distillate fuel support to BMCR. The maximum continuous rating of the boiler shall correspond to 105% of MCR.

The boiler main steam temperature shall be controlled by a combination of burner tilting, and desuperheating spray water, depending on the manufacturer's standards. If temperature control for the main steam is provided by desuperheating spray, the desuperheaters shall be located between the primary and secondary superheaters and the secondary and tertiary superheaters.

The reheat steam temperature shall be controlled by tilting burners. The secondary method of hot reheat steam temperature control shall be with desuperheating water spray.

The use of TP 347H shall not be allowed for the finishing superheater or reheater tubing. The use of TP 347 HFG, 304H, Super 304H, NF709, or HR3C shall be acceptable for the finishing superheat sections with internal shot peening for exfoliation resistance enhancement. The superheater and reheater layout shall provide sufficient mixing of flows to ensure uniform temperatures at all loads, and shall provide for thermal expansion of headers, tubes, and supports. All sections shall be accessible for cleaning, inspection, maintenance and removal of assemblies.

The furnace water walls shall use a membrane configuration with vertical (rifled) tubing.

The boiler shall be designed to operate satisfactorily over the full load range with the coal properties shown in Table 2-2 Rev C in Section 2.3 (Fuel) of the Employer's Requirements and the coal list included in Attachment B3. The boiler manufacturer shall design the boiler and the related auxiliaries to operate reliably with high moisture coal (up to 30 percent moisture by weight, as received) over the full load range of operation. The coal moisture content shall be considered in selecting the pulverizer design. In addition, the boiler related fans (including seal air fans) shall handle the additional moisture. The air heater design shall account for the need to provide hot primary air to the pulverizers to dry the wetter coal.

The furnace shall be sized to ensure complete combustion of the fuel within the furnace proper without flame impingement on waterwall, superheater, or reheater elements. A division wall shall not be used in the furnace. All radiant waterwall surfaces shall be installed in the walls, roof, or floor of the furnace without the use of water-cooled surfaces projecting into the furnace. Steam-cooled platens projecting into the furnace shall be widely spaced and designed for ease of cleaning during operation. Adequate space will be provided for full sootblower coverage of all sides of the furnace. Sufficient space for maintenance of sootblowers shall be provided.

The boiler heat input per plan area, burner zone heat release (tangential or front and/or rear fired design), heat available to the furnace per square meter of effective projected radiant surface, maximum gas temperature entering close spaced platens, and maximum gas velocity shall be selected by the Contractor to ensure stable and reliable operation with respect to slagging, fouling and erosion of the boiler over the range of fuels proposed at Boiler Maximum Continuous Rating (BMCR, normal pressure conditions), with carbon in the fly ash being no more than 5 percent (5%) for non-local coals, and no more than 3% for local coals with ash content greater than 2% on as received basis. The furnace exit gas temperature shall be less than the minimum initial deformation ash fusion temperature in a reducing environment for any fuel throughout the load range.

The Contractor shall provide flue gas velocity calculations during detailed design of the boiler back-pass, to confirm that high gas velocity zones have been eliminated with permanently fixed structures using material selection and design appropriate for the intended service. The Contractor shall prevent tube bank oscillations and noise using suitable measures, such as gas path partitioning. Gas velocities throughout the boiler shall not exceed 18 m/s.

Contractor shall provide an acoustic tube leak detection system complete with, amongst others, detectors, electronics and control room indicators.

Amongst other special tools which may be specifically required for the Plant and equipment maintenance, the Contractor shall provide a power operated cradle for furnace maintenance. The design of the maintenance cradle is such that the parts can pass through the furnace inspection/man hole and be fully assembled for use inside the furnace.

Furnace materials in the high heat zones shall be appropriate for the predicted duty and intended service.

A circulation analysis shall be developed to ensure proper design of water-wall tubing, headers, downcomers and feeder/releaser tubes. Results of the circulation analysis shall be shared with the Employer.

Superheater and reheater tubes shall be seamless. The economizer shall be of the non-steaming type and the economizer tubes shall be seamless and shall be bare, not finned. The Contractor shall demonstrate during detailed design that the furnace material selection is appropriate for the predicted duty and intended service, taking into consideration the heat flux and potential of flow assisted corrosion (FAC).

The boiler system (including the furnace, and all flues and ductwork from the air inlet to the stack) shall be designed for protection against negative and positive pressure per NFPA 85 requirements. The planned operation of the unit is base loaded operation at full load. However, the fatigue analysis of the boiler pressure parts and other related equipment shall be based on operation of a cycling unit (full load for 10 to 16 hours per day and reduced load corresponding to minimijim load without the use of supplementary fuel the remainder of the day).

A boiler filling system is also to be provided.

Vessels in once-through boilers which have the function of separating water and steam during startup and low load operation, shall be designed so that the separating ability of the vessel and internal components is adequate at all loads from ignition to complete once through operation with no water emitted to the super heaters.

Boiler water recirculating pump (BWCP) is to be provided. The pump shall be sized to adequately handle the maximum required flow during start-up and circulation, considering the prevailing conditions of pressure and temperature in normal and all possible transient conditions. The pump shall be installed in such a way as to have adequate access for maintenance, repair, removal and adjustment.

The Contractor shall install a minimum of 8 oxygen probes in the boiler backpass flue gas space in equal areas between the economizer and air heater.

The Contractor shall provide the system for short and long term preservation of the boiler and associated plant and equipment. The Contractor shall ensure that the wet storage regime is compatible with his installed dosing equipment.

Infrared (IR) temperature probes shall be provided to measure the temperature of flue gas at the furnace exit.

A complete soot blower system shall be furnished to effectively remove all ash, in its various forms, from the heat transfer surfaces of the complete boiler unit. Soot blowers shall be electric motor driven units. A means of manually rotating and retracting the soot blowers shall be provided. All elements shall be constructed of seamless tubing with one end closed. A soot blowing system for removal of ash deposits shall be provided with programmable type controls incorporating variable group selection of individual or groups of blowers as well as heat flux sensors for furnace tube temperature monitoring. The soot blowing steam header shall be equipped with an automatic drain to remove condensate. The soot blowing medium shall be steam for convective and superheat surface and furnace walls. Each sootblower shall be equipped with isolation valves to enable maintenance while the system is online.

The type of air heater and boiler auxiliary equipment shall be dependent upon vendor selection and associated OEM standards. Air pre-heater design and material selection shall be appropriate to mitigate acid dew-point corrosion issues.

Furnace design pressure shall be ± 533 mm w.g.

All combustion air ductwork shall be designed for minus 203 mm w.g. and a positive pressure equivalent to the maximum transient pressure that can be developed at any location during operation (including a Master Fuel Trip) plus a margin.

All flue gas ductwork up to the inlet of the ID fan shall be designed for plus 203 mm w.g. and a negative pressure equivalent to a worst transient negative pressure that can be developed at any location during operation (including a Master Fuel Trip) plus a margin.

In no case shall the design pressure be less than that required by NFPA 85.

The boiler management control system shall be in the DCS processors that comply with NFPA 85 requirements. The boiler shall be supplied with the instrumentation required for the burner management control system, including main steam pressure and temperature transmitters, reheat pressure and temperature transmitters, combustion airflow and temperature transmitters, and zirconium oxide cell type combustion gas O_2 analyzer. Finishing main steam temperature control shall be accomplished using a desuperheating spray water control valve. Steam supply to the soot blower system shall be regulated with an air actuated control valve. An array of boiler tube metal temperate thermocouples shall be provided and connected to the DCS for operator information.

An 813 mm wide by 1397 mm high water cooled access door shall be provided on each side of the lower furnace. 457 mm wide by 610 mm high refractory lined access doors shall be provided to permit access to superheater, reheater, between the arch tubes and casing, between economizer banks, and in any other areas requiring access. All doors shall be hinged and pressure tight. Sight holes shall be provided to give a clear view of each burner row and side, front and rear walls in the vicinity of burners, as well as the ash hopper slopes. Any other area requiring inspection during operation shall be similarly fitted. These holes shall be situated not higher than 1.5 m from the adjacent floor or gallery levels with easy access without building of scaffolding.

Chapter Three Properties of VRE Generators

第三章 VRE发电机的特性

原文

The political debate regarding VRE integration is often complicated by the technical complexities of the subject. The consequences of VRE taking an increased share of energy supply are, as a rule, system-specific. VRE integration is interactive: properties specific to VRE meet those of power systems, and their interaction determines adaptation effects and ultimate impacts. However, a limited number of properties, both of VRE and power systems, largely shape integration effects. ⑧ They are analyzed by asking whether they are a transition phenomenon, which results from adding a new technology to an existing system, or if effects are more fundamental.

① VRE power plants are linked to the power system more indirectly via power electronics and have less or no spinning mass (inertia); VRE sources are thus said to be non-synchronous generation technologies. This property may require changes to how system stability is

译文

关于可变的可再生能源(variable renewable energy, VRE)整合的政治辩论常常由于学科的技术复杂性而变得复杂。VRE在能源供应中占据的比例不断增加,造成的后果是影响了系统特性。VRE整合是交互式的:VRE特有的属性符合电力系统的特性,而且它们的交互决定了适应效应和最终影响。然而,VRE和电力系统的有限数量的特性在很大程度上决定了整合效应。分析这些特性就是探讨它们是否是由于在现有系统中添加新技术而产生的一种过渡现象,或探讨是否这些整合效应是更基本的。

VRE发电厂通过电力电子装置间接地与电力系统相连,并且具有较小的或不具有自旋质量(惯性);因此,VRE电源被称为非同步发电技术。这种特性需要改变,以保证电力系统的稳定性,尤其是在VRE发电比例

ensured, especially during periods of high shares of VRE in power generation. A detailed, system-specific integration study is necessary for a thorough understanding of integration impacts. Broadly speaking, if local concentrations ("hotspots") are avoided, VRE deployment has a negligible impact at shares of around 2% to 3% of annual generation. ② Apart from small island systems, shares of 5% to 10% in annual generation will not lead to technical integration challenges if operations are adapted and VRE deployment patterns are well co-coordinated. The political debate regarding VRE integration is often complicated by the technical complexities of the subject. The consequences of VRE taking an increased share of energy supply are, as a rule, system-specific. This part seeks to organize the various impacts of VRE deployment into a more coherent framework.

较高的时期。一个详细的、系统特定的整合研究对于全面理解整合的影响是必要的。一般来说,如果能避免本地集中("热点"),VRE 的部署在占年发电量 2%～3%时产生的影响可以忽略不计。除小岛系统外,如果采取的运行方式合适而且 VRE 部署模式能很好地协调,那么年发电量比例为 5%～10%也不会导致技术上的整合挑战。关于 VRE 整合的政治辩论往往因这一学科的技术复杂性而变得复杂。VRE 在能源供应中占据的比例不断增加,造成的后果与系统特性相关。本部分内容旨在将 VRE 部署的各种影响因素组织成一个更连贯的框架。

1 Properties of VRE Generators

VRE generators have a number of specific characteristics that affect their contribution to power system operation and investment. Knowledge of these characteristics continues to evolve. Without implying any order in importance, VRE generators are:

- low short-run cost, i.e. once

1 VRE 发电机的特性

VRE 发电机具有许多特有的性质,这些特性影响了其对电力系统运行和投资的贡献。关于这些特性的知识在不断发展。不按照任何重要性次序,VRE 发电机的特性有:

- 短期成本低,即一旦安

installed they can generate power at very little cost—their short-run costs are close to zero.

• variable, i.e. available power output fluctuates with availability of resource (wind or sun).

• uncertain, i.e. the availability of resource can only be predicted with high accuracy in the short term.

• location-constrained, i.e. resource is not equally good in all locations and cannot be transported.

• modular, i.e. ③ the scale of an individual VRE production unit (wind turbine, solar panel) is much smaller than fossil, nuclear and larger hydro generators.

• non-synchronous, i.e. VRE plants connect to the grid via power electronics, in contrast to large conventional generators, which are synchronized to the grid and therefore react in a co-coordinated fashion to changes in the grid.

装，VRE只需要极低的成本就能发电——它们的短期成本接近于零。

• 多变性，即可用的电能产量会随着可用的资源（风能或太阳能）可利用性波动。

• 不确定性，即只能在短期内以高精度预测资源的可利用性。

• 地理位置局限性，即所有的地理位置的资源并不是同样良好，并且不可输送。

• 模块化，即单独的VRE发电机组（风轮机，太阳能电池板）规模远小于化石燃料、核燃料以及大型水力发电机。

• 非同步，即VRE发电厂通过电力电子装置与电网相连，与之相比，大型传统机组是同步并网，因此是以协同协调的方式对电网变化做出反应。

2 Low Short-run Cost

The first property (low short-run cost) is not a technical property but has important impacts on electricity markets and is therefore included in the above list. However, contrary to the other five, there is no technical integration issue associated with it. ⑨ The above

2 低短期运行成本

低短期运行成本不是一个技术特性，还对电力市场有重要的影响，因此会包括到上述列表里。然而，与其他五项相反，低短期运行成本没有与之相关的技术整合问题。上述六个特性产生目前众多可见的VRE整合

properties contribute to all currently observed integration drive impacts of VRE. The penetration level at which they become relevant depends on system-specific circumstances.

的拉动作用。它们可引起相关影响的渗透水平取决于系统特定的情况。

④Once built, wind power and solar PV provide electricity practically for free. However, in the absence of demand growth or power plant ⑤retirement, the integration of this additional power generation is only possible by reducing the market share of ⑤incumbents. VRE deployment takes place in many member countries of the Organization for Economic Co-operation and Development (OECD) that have such stable power systems, characterized by adequate generation capacity and slow demand growth.

风电厂和光伏电厂一旦建成,风能和太阳能几乎可以免费发电。然而在没有负荷需求增长和电厂停止运行的情况下,只有通过减少电力生产企业所占市场份额才能实现对这种额外发电的接入。经济与合作与发展组织(经合组织)中的成员国许多具有稳定电力系统,具备足够发电能力并且缓慢增长,出现大量可再生能源部署。

In a purely competitive and fossil fuel dominated environment, such displacement effects are common and frequent. The merit-order of fossil fuel fired power plants has always been affected by variations in resource prices. For example, lower gas prices in the United States have recently boosted gas generation at the expense of coal. The low short-run costs imply that once VRE generation is built, it is likely to be among the first technologies in the merit-order. It will therefore displace more

在完全竞争的市场和化石能源主导的环境下,大量可再生能源取代传统能源十分常见。资源价格波动影响化石燃料电厂发电的优先次序。例如,最近美国天然气价格降低,从而以牺牲煤炭来提高天然气发电量。低短期成本意味着大量 VRE 电厂一旦建成,它的技术优先等级最高。因此,可再生能源发电将会取代燃气发电、燃煤发电这类通常以燃料成本和二氧化碳排放成本组合为主的发电成本,因

costly generation — usually gas, or coal, whichever has the highest short-run cost (generally dominated by a combination of fuel and carbon dioxide (CO_2) emission costs).

为燃气发电和燃煤发电短期运行成本最高。

⑥Based purely on plant economics, VRE generators would be expected to bid no lower than at a very low, positive price, reflecting their very low short-run cost. They would not be expected to bid below zero. However, support policies often contain a performance-based element, i.e. they remunerate based on generated energy (e.g. feed-in tariffs, feed-in premiums, tradable green certificates or production-based tax incentives). In cases where VRE generators offer their generation directly to the market, such remuneration may create an incentive for VRE plant owners to bid below their short-run costs, because they receive revenues on top of achieved market prices. Hence, bids may be below zero (minimum bids are likely to equal short-run cost minus the value of support payments).

如果单纯基于电厂经济学，VRE发电机组应该不会在低于零的价格中标，这反映了新能源机组极低的短期运行成本。然而政策支持通常会考虑绩效因素，即根据发电量来获得报酬，例如上网电价、回购电价、可交易的绿色证书和基于生产的税收优惠等。如果VRE发电直接进入电力市场，这样的报酬可能会刺激新能源电厂所有者出价低于短期运行成本，因为他们基于市场价格获得收益。因此，最低出价等于短期运行成本减去维持成本，因此出价可能会低于零。

⑥Depending on the policy context, VRE generators may also enjoy priority dispatch. This means that they are treated as must-run units, i.e. they are allowed to generate whenever wind and sun are available. This is achieved via different mechanisms in different power

由于政策因素，VRE发电可以享受优先调度。这意味着可再生能源机组是必要运行机组单元，也就是说在不同的电力系统中，通过不同的机制允许VRE机组在有风和阳光的条件下发电。例如德国输电系统运

systems. For example, German transmission system operators (TSOs) generally offer all renewable energy produced under a feed-in tariff at the minimum price on the European Power Exchange (EPEX). This price is currently EUR-3 000 per megawatt hour (/MWh). If prices clear below EUR-150/MWh, TSOs resubmit their bids at a randomly chosen value between EUR -150/MWh and EUR -350/MWh to contain negative prices. ⑦ Where VRE generators have priority dispatch, their operation can run independent of any market price signal. This can lead to more pronounced negative prices.

营商通常会以最低上网电价向欧洲电力交易所提供所有可再生能源发出的电能。目前最低上网电价是每兆瓦时 3000 欧元。如果上网电价低于 150 欧元每兆瓦时,德国输电系统运营商会重新提交从 150 欧元每兆瓦时到 350 欧元每兆瓦时的随机报价。在欧洲电力交易所,因为 VRE 机组享有优先调度权,所以运营商的操作独立于市场上任何价格信号。这可能导致明显的负价格。

《术语表》

VRE	可变的可再生能源
spinning mass	自旋质量
short-run cost	低短期成本
synchronous	同步
penetration level	准入容量
displacement effect	替代效应
merit-order	排队,优先次序
feed-in tariff	上网电价
feed-in premium	价格溢价,也叫回购电价
TSO	输电系统运营商
EPEX	欧洲电力交易所
priority dispatch	优先调度

翻译难点讲评

① 本句采用直译策略，主要按照文字的字面意思直接翻译过来，目的是为了表意义清楚，毕竟作者希望他的读者对象的认知语境存在相似性。比如VRE没有具体化为“可变的可再生能源”，保留了缩写；两个术语“spinning mass”和“non-synchronous generation technology”也采用的是术语直译策略，再现了源语术语的形式，分别翻译为“自旋质量”和“非同步发电技术”。全句的语序基本保留了原句的形式，但要求得“最自然、最贴切”的译文，原文与译文无法求得完全对等，要在汉语表现力的范围中，保存原文的风格，表现原文的意义。比如，“indirectly”这个副词作状语，在英语中可以放在谓语动词的后面，但是汉语中需要提前到动词之前，“have”这个动词原文为了表达简洁只使用一次，所以没有用“have less or have no”，但是译文中重复使用“具有”，两次分别译为“具有较小的”和“不具有”。

② 本句为英语长句，使用了重组法。由于英汉两种语言在结构、语法和表达方式上存在很大差异，如果将句子直译或顺译出来，那么肯定不符合译入语的表达习惯和思维方式。这时，我们就要大胆地打破原文句子结构，在不改变原文句意和中心思想的情况下，根据目的语的表达习惯和思维方式重新整合句子，不仅力求译出源语的准确句意，而且符合读者的接受习惯。原句中的if条件句放在了主句的后面，而汉语的表达习惯是先说条件，所以译文把条件句调整到主句的前面。另外语境因素，是所有翻译活动中必须要考虑到的因素。本句的上面一句是“Broadly speaking, if local concentrations（‘hotspots’）are avoided, VRE deployment has a negligible impact at shares of around 2% to 3% of annual generation.”。所以，本句中的“if operations are adapted”，结合上文语境，应该内容有省略，需要补充说明是什么“operation”，以及被采取后的结果如何，因此译文对句子成分进行了调整，并增译了省略成分，译为“如果采取的运行方式合适”。可以看出，在应用重组法中，将语境因素也考虑进翻译过程，可以使译文变得更流畅自然，更明晰化作者的写作意图。

③ 由于英汉两种语言表达习惯不同，翻译时如果一字不漏地照搬译出，往往会显得累赘或不合行文习惯，甚至产生歧义，采取省译这种译法可以使译文言简意赅，流畅表达（龚明星，2011：136-137）。原句中的“scale”和“generators”是多支共干结构的共有成分，如果在译文中风轮机、太阳能

电池板后面都加上"规模",在"化石燃料、核燃料以及大型水力"后面都加上"发电机",不但烦琐,且无必要。可见多支共干结构中的共有成分在译成汉语时为了避免太过啰唆或造成太大的累赘有时可将其省略。

④ 由于英汉两种语言词法和句法的差异,使其表达方式不尽相同。翻译时为了使译文合乎汉语的习惯和表达规律,意思更加明确清晰,从意义、修辞和句法上考虑,根据上下文按照某些原则在译文中适当增添一些虽无其词但有其意的词,从而达到源语与译入语的和谐与完美。比如此句"Once built, wind power and solar PV provide electricity practically for free."不仔细分析前半句的逻辑主语,可翻译成"一旦建成,风能和太阳能几乎可以免费发电。"但是根据上下文,都在探讨新能源发电厂,此句的逻辑主语可见应该是风电厂和光伏电厂,需要增译出来。

⑤ 科技术语翻译要讲科技,似乎是科技领域的"天经地义"。但这里的两个单词"retirement"和"incumbent"都带有隐喻色彩。查询字典得到"Retirement is the time when a worker retires."意思是"退休",而"An incumbent is someone who holds an official post at a particular time."意思是"在职者"。在特定语境中,对科技英语中暗含隐喻意义的词,则宜译隐喻的喻体,而不译相应隐喻的本体。所以,在科技英语中对于含有隐喻的词汇的翻译方法除了直译外,更重要的还是灵活使用好意译及解释加补译等手段。"retirement"意译为"停止运行","incumbent"解释并补充说明为"电力生产企业"。

⑥ 英语中的分词短语一般是作状语的,可表达各类型的状语,例如伴随、原因、条件等状语。因此,汉译此类句子时应增译某些关联词,使译文忠实地表达原文的意思与风格并使译文合乎表达习惯。这里的两个句子中分别有"Based purely on plant economics"和"Depending on the policy context",分别表达的条件状语和原因状语,在翻译成汉语的时候,要补充关联词"如果"和"因为",以达到承上启下连接句子作用。

⑦ 本句中的从句引导词"where"如何处理非常重要,直接影响到译文意义表达的准确性。直译成为地点状语从句是不合适的,而如果根据上下文翻译成它的所指对象"在欧洲电力交易所"就会明晰很多。再仔细查看可发现,原文从句和主句的关系实际上是因果关系,可增译关联词"因为……所以",最后翻译成"在欧洲电力交易所,因为 VRE 机组享有优先调度权,所以运营商的操作独立于市场上任何价格信号。"

翻译理论及应用讲评:生态翻译学

从生态学角度研究翻译是胡庚申首先开辟的研究路径。其是在翻译适应选择论的基础上发展起来的,起步于2001年胡庚申在香港浸会大学的讲座“From the Darwinian Principle of Adapation and Selection to Translation Studies”和第三届亚洲翻译家论坛论文“初探翻译适应选择论”。胡庚申在2004年出版了《翻译适应选择论》专著,以及2013年的国家社科基金研究成果专著《生态翻译学·建构与诠释》,从而确立了一套生态翻译学的理论体系。该理论有九大核心理论视角:生态范式,关联序列链,生态理性,译有所为,翻译生态环境,译者中心,适应/选择,三维转换以及事后追惩。其中翻译生态环境包括原文、源语和译语系统等在内的各种因素,是译者和译文生存的总体环境,影响到译文内容、风格和选词造句,是该理论的核心概念之一。翻译适应/选择论意味着译者对翻译生态环境的适应中有选择,即适应性选择,在对译文的选择中有适应,即选择性适应,这一过程交替循环进行。适应的目的是求存,适应的手段是优化选择,而选择的法则是汰弱留强,在选择性适应和适应性选择的过程中追求整合适应选择度最高的翻译。三维转换是生态翻译学基础理论在多维度适应与适应性选择的原则之下提出的翻译方法,即语言维(语义形式的转换)、文化维(文化内涵的传递与阐释)、交际维(双语交际意图在译文中的体现)的适应性选择转换,简称三维转换。因此,译者翻译时应关注语言差异、文化差异、交际意图,并作为选择转换的侧重考虑因素,致力于这三个维度的平衡和和谐。下面结合本章文本和译文举例说明生态翻译学的多维度适应与适应性选择,以及三维转换在科技翻译中的应用。

例⑧“They are analyzed by asking whether they are a transition phenomenon, which results from adding a new technology to an existing system, or if effects are more fundamental.”。译文为“分析这些特性就是探讨它们是否是由于在现有系统中添加新技术而产生的一种过渡现象,或探讨是否这些整合效应是更基本的。”此句体现了语言维的转换,译者在主语的人称物称转换,句型的主动被动转换上做出选择。英汉语的语言表达偏好有极大的差别。中西方思维方式的差异对语言表达方式有明显的影响。首先,原句中的代词“they”所指内容需要明晰化,而且原文的被动需要转换为主动。这一代词还原转换和主被动转换都符合了汉语的表达习惯。原句介词

+动名词的结构“by asking”要相应省略介词，并将动名词转换为动词词性。英汉两种语言在词性使用的偏好上有不同，前者大量使用名词，既可表达抽象的概念，也可表达动作特征。英译汉时，词性转换可增加译语语言可读性和语义明确性，也可让句子结构更加合理。比如此句译文连用两个“探讨”来对应表达“asking”的意义，就避免了把介词短语译为方式状语所带来的句子冗长和头重脚轻的问题。

例⑤原文“However, in the absence of demand growth or power plant retirement, the integration of this additional power generation is only possible by reducing the market share of incumbents.”译文为：“然而在没有负荷需求增长和电厂停止运行的情况下，只有通过减少电力生产企业所占市场份额才能实现对这种额外发电的接入。”该句体现了文化维的转换。原句的“retirement”和“incumbent”都是带隐喻色彩的词汇，其增加了文字的活泼生动性，但也为译者提出了翻译难题。此处，文化维的转换就是把隐喻词汇的真正含义表达出来，对该词的基本内涵意义和概念意义适当取舍。所以不能翻译为“退休”和“在任者”。为了维持翻译生态环境平衡和和谐，为了把作者意图传达给读者，在本文所处的环境中，应该分别把它们翻译为“停止运行”和“电力生产企业”。

例⑨原文“The above properties contribute to all currently observed integration drive impacts of VRE. The penetration level at which they become relevant depends on system-specific circumstances.”译文为：“上述六个特性产生目前众多可见的VRE整合的拉动作用。它们可引起相关影响的渗透水平取决于系统特定的情况。”此例体现了交际维的适应性转换。该处原文的意思是VRE的特性对它整合入电网造成冲击影响，相关影响水平与系统特征相关。如果按照原文形式翻译，第一句表达没有不妥，关键问题会出现在第二句。因为两句的主语对象不一致，会引起信息传递的障碍。汉语是一种主题突显（topic-prominent）的语言，与英语的主语突显（subject-prominent）的特征不同。为了在译文中体现出主题相关和陈述对象的一致，有必要在第二句增加主语“它们”，与上一句中的主语对象相同。此外，尽管与上一句的内容有所重复，还是有必要增加“引起影响的”，以配合新增加的主语“它们”。转换后的译文可最大限度消除读者的理解困难。反之，如果第二句也按照原句语序翻译，译文没有了与前一句的衔接，会显得不自然，不连贯。

补充阅读:Generator

A generator has an input shaft that is driven by a "prime mover" such as a diesel engine, or a steam, wind, or water (hydro) turbine. The prime mover supplies mechanical energy to the generator input shaft. Inside the generator, the input shaft turns a "rotor" that creates a rotating magnetic field. This field crosses "stator" (fixed or static) "windings" (wire coils) in the generator housing. These windings are connected to the generator output. The changing magnetic field creates a voltage on the generator output by the process of magnetic induction. When the generator output is connected (through a transmission system) to a load, currents flow that carry electrical energy to the load.

There are two main types of generator in common use in power systems: synchronous generators and induction generators. Most generators are synchronous generators. Electric motors are available in corresponding forms; however, synchronous motors are rare and used only in special applications.

1. Synchronous Machines

A synchronous machine, as indicated by its name, requires the rotor to spin at a fixed speed relative to the frequency of the power system. The rotor spinning frequency can be equal to that of the power system (eg. 3000 RPM), or reduced to the frequency, while the additional rotor coils are called "poles (salient or non-salient)" and the prime mover can spin at other speeds rotor.

The rotor has a winding that is powered by a steady (direct current - DC) power supply. The constant current forms a magnetic field, which rotates with the rotor. A synchronous generator can control its output voltage by changing the amount of current fed into the rotor winding. Historically, this has generally been the cheapest way to provide fast voltage control on power systems, and so almost all generators (and all large generators) have been of the synchronous type.

2. Induction Machines

Induction machines are not speed-locked to the system frequency, although in normal operation they operate within a small speed range. Their chief advantage is that they are mechanically simpler, and cheaper to manufacture. However, they have adverse effects on power system performance (compared with synchronous machines or other types of load). Two common induction machines are the "squirrel cage" and the "Doubly Fed Induction Generator (DFIG)".

The squirrel cage induction machine is named because of the use of a cylindrical rotor cage inside the rotor, which resembles a cage on which squirrels exercise, or a mouse wheel.

The rotor does not require a power supply to maintain a magnetic field. Instead the energized stator windings induce currents to flow along the rotating rotor bars, forming a magnetic field. The rotor is very tolerant to changes in speed because the magnetic field present on the rotor is not held in place, as it would by an external power supply in a synchronous machine.

A Doubly Fed Induction Generator (DFIG) operates like a combination of a normal induction machine (such as the Squirrel Cage) and a synchronous machine. Inverter fed three-phase windings on the rotor allows the magnetomotive force (mmf) of the rotor to rotate independently of the mechanical rotation. With a suitable control system for the inverter the rotor mmf stays in synchronism with the power system frequency while the rotor speed varies.

A DFIG is very useful in wind generation, as the blades can spin at full range of speeds. The aerodynamic efficiency of wind turbines is greatest at a particular rotation speed, and the optimal speed varies when the wind speed changes. Therefore, variable speed operation maximizes the efficiency of the wind capture. In addition, this type of generator can help to control system voltage (unlike a squirrel cage induction machine). Usually the inverter is only small (relative to the machine size) and so the speed range and amount of voltage support are both limited.

3. Inverter Connected

An inverter is an electronic device that can be used to connect generators to the network. The inverter outputs an alternating current (AC), from a direct current (DC) input or from an input that is oscillating at a different frequency to the power system. This allows a source that produces DC (such as a photovoltaic system) to be connected to the power system (which uses AC).

An inverter can also enable a synchronous machine to be operated at variable speed. Compared with a DFIQ a mechanically simpler arrangement can be used (by avoiding the use of high power slip-rings) so long as the inverter does not have a smaller rating than the main machine output.

Inverters can provide excellent voltage control, even when the energy source is unavailable.

4. Comment

In practice, most generators on power systems are synchronous (and regulate voltage), and most loads are induction motors (with the remainder being resistive or static). In principle, it would be possible to reverse this and have a power system with induction generators and synchronous motor loads (with the loads regulating voltage). It could even be argued that this is preferred, because the load voltage is of most interest and it is better that the regulation occurs near the load.

However, there are few generators, and there are very many loads. Generally, it is cheaper and more efficient to provide the expensive (and somewhat complicated) voltage control equipment on the small number of generators. In principle, it is not necessary to provide this control on every generator. There have been proposals in the technical literature to use large, induction generators; however, this has not eventuated in practice.

Many early wind generators used induction machines. These are cheaper and, importantly, require a lighter, mechanically simpler, and less maintenance-intensive machine in the turbine nacelle. This was acceptable while the machines and wind farms were relatively small but they are now approaching or exceeding the size of conventional power plants. Given the circumstances outlined above, induction generators are increasingly being required to install supplementary voltage control equipment to overcome the inherent limitations of the generators.

Chapter Four Hanford Site Radioactive Tank Waste Treatment and Immobilization

第四章 汉福德场区罐体核废物固化处理

原文

Forty years of plutonium production at the Hanford Site has yielded a challenging nuclear waste legacy—approximately 56 million gallons of radioactive and chemical wastes stored in 177 underground tanks (tank farms) located on Hanford's Central Plateau. The mission of the U.S. Department of Energy (DOE) Office of River Protection (ORP) is to address the risks posed by this tank waste through immobilization of the waste, and the ultimate closure of the tanks and decommissioning of the treatment facilities. While there are no immediate risks to the Hanford workforce, the public, or the accessible environment from leaking tanks, DOE remains steadfastly focused on treating Hanford's tank waste as safely and expeditiously as possible.

译文

汉福德场区40年的钚制造产生了具有挑战性的核废物遗留——大约5600万加仑的放射性和化学废物储存在位于汉福德中部高原的177个地下储罐(储罐场)中。美国能源部河流保护办公室(DOE ORP)的使命是通过固化这些核废物,以及最终关闭这些储罐并关停处理设施,来应对这些罐体废物所带来的风险。虽然汉福德的工人、公众或周围的环境还未因泄漏的罐体核废物而受到直接的威胁,但美国能源部仍然坚定地致力于尽可能安全和迅速地处理汉福德的罐体核废物。

The tank waste is currently stored in aging single-shell (SST) and double-shell tanks (DST). The liquid portion of the

目前,罐体废物储存在老化的单壳罐(SST)和双壳罐(DST)中。核废物的液体部分最有可

waste is the portion most likely to leak from the tanks. 7 DOE took steps beginning in the 1980s to mitigate this risk by transferring all pumpable liquids from the older single-shell tanks to newer double-shell tanks. The next step is to immobilize this waste.

能从储罐泄漏的。从 20 世纪 80 年代开始，美国能源部采取了一些措施来降低这种风险，将所有可泵送液体从旧的单壳罐转移到新的双壳罐内。下一步措施是将这些核废物固化起来。

Immobilization will occur in ① the Waste Treatment and Immobilization Plant (WTP). The WTP is a highly complex nuclear and chemical processing facility with many first-of-a-kind technology applications. The tank waste at Hanford is also the most complex and heterogeneous radioactive tank waste in the United States. ⑥ The complexity of both the waste itself as well as the WTP facilities has led to difficult, and to date, unresolved technical issues for the portions of the facility (primarily the Pretreatment [PT] Facility and to a lesser extent the High-Level Waste [HLW] Facility) that will process the solid portions of the waste. Because the current design of WTP anticipates that all waste will be processed through the PT Facility, immobilization of any waste could not occur per the current plan until the technical issues involving the PT Facility are resolved. Therefore, an alternative approach for immobilizing waste as soon as practicable, while simultaneously resolving the remaining technical challenges, has been identified.

固化将发生在废物处理与固化厂（WTP）中。废物处理与固化装置是一个高度复杂的核和化学处理设施，应用了许多最为先进的技术。汉福德的罐体核废物也是在美国最复杂和最混杂的放射性罐体核废物。核废物本身和废物处理与固化厂设施复杂，以至于部分关于废物固化处理设施（主要是预处理设施和较小范围内的高放射性核废物处理设施）的技术难题至今尚未解决。由于废物处理与固化厂目前的设计是预计所有废物都将通过预处理设施处理，所以在涉及预处理设施的技术问题得到解决之前，任何废物的固化都无法按照当前的计划进行。因此，制定了一个尽量可行的，同时可解决其余技术挑战的核废物固化替代方案。

The technical issues will take time to resolve, but DOE has assembled dedicated teams of DOE, contractor, national laboratory, and industry experts and is devoting significant resources to resolve these issues. At the same time, by adopting a ① DFLAW option in which the waste bypasses the PT Facility, waste immobilization can begin years earlier than if we wait until all technical issues are resolved and the Pretreatment Facility is completed.

技术问题的解决需要时间，但美国能源部已经召集了能源部、承包商、国家实验室和行业专家成立的专门小组，并投入了大量资源来解决这些问题。同时，通过采用核废物绕过预处理设施的直接进料低放射性废物，(direct-feed low-activity waste, DFLAW）方案，可以比等待所有技术问题的解决和预处理设施的完成，更早几年开始固化废物。

This document is not a proposal, but rather a framework for discussion as DOE and the State of Washington seek to resolve concerns regarding completion of the waste treatment mission. Viewed as a whole, this Framework describes an approach that would allow for immobilization of tank waste to begin as early as practicable without waiting for completion of work to resolve the technical issues associated with the PT and HLW Facilities. For each of the waste streams described in more details, this framework identifies potential waste treatment options, based on a ⑤ combination of previous alternatives analyses, external reviews, testing, and ongoing analyses. Where possible, the document identifies a preferred alternative.

这个文件不是一项建议，而是考虑到能源部和华盛顿州寻求解决有关完成废物处理任务的关切，提出的一个讨论的框架。从整体上看，这一框架描述了一种办法，可使罐体废物的固化尽早开始，而无须等待与预处理设施和高放射性核废料处理(HLW)设施有关的技术问题的解决。对于更详细描述的各个核废物流，该框架综合先前的备选方案分析、外部审查、测试和正在进行方案的分析，以此为基础确定了可行的核废物处理方案。在可能的情况下，该文件确定了一种首选的替代方案。

1 Current System Design

1 当前系统设计

The WTP consists of five facilities/complexes: (1) the Analytical Laboratory

废物处理和固化厂由五个设施组成：(1) 分析实验室

(LAB), (2) Balance of Facilities (BOF), (3) LAW Facility, (4) HLW Facility, and (5) PT Facility. The WTP is being designed to process the tank farm waste during a roughly 40-year period. The current design requires waste to be processed through the PT Facility, where it will be separated into a ③ low-activity waste stream to be vitrified in the LAW Facility and a high-level waste stream to be vitrified in the HLW Facility. The LAB and BOF support these ② vitrification activities.

(LAB),(2)辅助设施(BOF),(3)低放射性核废物设施(LAW),(4)高放射性核废物设施(HLW)和(5)预处理设施(PT)。废物处理和固化装置的设计目的是为了在大约40年的时间内处理油罐区的废物。当前的设计是,先将废物通过预处理设施进行处理,在预处理设施中废物将被分离成低放射性废物流,然后在低放射性废物设施中玻璃固化,而高放射性废物流将在高放射性废物处理设施中玻璃固化。分析实验室和平衡设施会促进这些玻璃固化过程。

The LAW Facility, BOF, and the LAB — collectively referred to as the LBL — are nearest to completion and do not have any significant remaining technical issues. As described in more detail in the following sections, technical issues associated largely with mixing in the vessels of primarily the PT Facility, and to a lesser extent the HLW Facility, ④ have caused construction of the PT Facility to be suspended and construction of the HLW Facility to be slowed.

低放射性废物设施、辅助设施和分析实验室(统称为LBL)已接近完工,不存在太大的技术问题。以下主要描述的是与预处理设施容器中的混合有关技术问题,以及少部分与高放射性废物处理设施中的混合有关的技术问题,这些技术问题导致预处理设施的建造工程暂停以及高放射性废物处理设施的建造进度放缓。

2 Addressing Technical Risks and Challenges

2 应对技术的风险和挑战

Hanford tanks contain a complex and diverse mix of radioactive and chemical

汉福德核废物储料罐场含有复杂多样的由放射性废物和

waste in the form of sludge, salts, and liquids, necessitating a variety of unique waste retrieval and treatment methods. The uncertainty and diversity of the physical and chemical properties of the 56 million gallons of waste make the mission uniquely complex.

化学废物混合而成的污泥、盐和液体，需要各种独特的废物回收和处理方法。而且 5600 万加仑废物的物理、化学性质的不确定性和多样性使得任务变得异常复杂。

The underground tank farms at Hanford include 149 older SSTs that are decades past their design life. Some of these tanks are known or are assumed to have leaked, and some SSTs continue to slowly leak. ⑧Although there is no immediate health and safety risk posed by these leaks, addressing the long-term environmental concerns associated with the leaks requires a robust and sustainable strategy for waste retrieval, treatment, and disposal or long-term storage. This strategy, thus far, has involved transferring the pumpable liquid waste that posed the highest risk of leaking to the environment from the SSTs to 28 DSTs.

汉福德的地下储料罐场包括 149 个单壳储料罐，这些设备设计寿命数已经超过了十年。这些储料罐中的一些已经或者推测已经泄漏，有一些持续缓慢泄漏。虽然这些泄漏不会带来直接的健康和安全风险，但是要解决与泄漏相关的长期环境问题，需要能源部针对核废物回收、处理和处置或长期储存制定强有力的可持续战略。迄今为止，该战略已经包括将泄漏风险最高的可泵送液体废物从单壳罐转移到 28 个双壳罐。

As the design and construction of the WTP has progressed, a number of technical issues have emerged involving the tank farms, the WTP, and the interfaces between the two. As previously noted, the issues in WTP are primarily associated with the PT Facility and, to a lesser degree, the HLW Facility. However, because in the current design all waste

随着污水处理厂的设计和建造的进展，出现了一些涉及储料罐场、污水处理厂以及两者之间接口的技术问题。如前所述，废物处理和固定装置中的问题主要与预处理设施相关，其次与高放射性废物处理设施相关。然而，由于在当前的设计中，所有废物流经过预处理设施，因此

flows through the PT Facility, these technical issues impact ORP's overall ability to begin treating Hanford's tank waste.

这些技术问题影响了河流保护办公室处理汉福德地区废物的总体能力。

The WTP technical issues are centered on the ability of the PT Facility to mix and transfer HLW slurries with high solids concentrations and the adequacy of the piping and vessel designs in inaccessible black cells to support the WTP's 40-year operational life. Some of these issues involve uncertainties associated with the erosion and corrosion of piping and vessels, criticality, and ⑤ hydrogen generation in vessels.

废物处理和固化厂技术问题集中在预处理设施混合和输送高浓度浆料的能力,以及管道和容器设计妥善性,从而可以维持废物处理和固定装置 40 年的运行寿命。其中一些问题涉及管道和容器的腐蚀和侵蚀、临界、容器中氢气生成等相关不确定性因素。

《术语表》

radioactive tank waste	罐体核废物
single-shell/double-shell tanks(SST/DST)	单壳/双壳罐
waste treatment and immobilization	废物处理与固化
HLW	高放射性核废物
LAW	低放射性核废物
vitrification	玻璃化
criticality	临界

翻译难点讲评

① 作为一个关键术语,WTP 指的是核电厂运行过程中产生的高低中水平放射性核废料的处置与管理,它的英语全称是"the Waste Treatment and Immobilization Plant (WTP)",查阅知网翻译助手(dict.cnki.net),在化学、有机化工、环境科学与资源利用等学科领域"immobilization"多被翻译为"固定化"或者"固定"。但是通过网络搜索,发现在几篇有关核放射性废物处理和处置的文献中,作者在文章中用的是"固化",比如在"郭喜良,冯声涛. 低、中水平放射性废物水泥固化体性能检验的讨论[J]. 辐射防护通

讯，2007，27(3)：12-16.""范智文，郭喜良，冯声涛．放射性废物固化体抗浸出性快速测定方法探讨[J]．原子能科学技术，2007，41(5)：540-545.""刘志辉，冯声涛，程理．放射性焚烧灰稳定/固化技术进展[C]//放射性废物处理处置技术经验交流会．2005."。由此可以看出这个缩略词"WTP"可以翻译成"废物处理与固化厂"。而另外一个缩略词"DF-LAW"，可以在"祝疆．英汉核科学技术缩略语词典[M]．原子能出版社，1984."这本词典中找到，"DFHLW"整词形式"direct-feed high-level waste"，翻译为"直接进料高放废物"，"DFLAW"整词形式"direct-feed low-activity waste"，翻译为"直接进料低放废物 "。

② 这个纯技术术语的词"vitrification"，查柯林斯英汉双解大辞典，给出的解释是"the process or act of vitrifying or the state of being vitrified，玻璃化；呈玻璃状"。但是上下文中提到的都是对低放射性废物流和高放射性废物流通过预处理设施进行处理后再进一步的处置，而且正文都在说核废物的固化，那么这里只翻译成为"玻璃化"是否不能清晰表现出该术语的意义？带着这个问题搜寻网络，发现一篇文章："刘志辉，冯声涛，程理，等．焚烧灰造粒水泥玻璃固化配方研究[J]．辐射防护，2006，26(1)：17-23."，这篇文章中作者使用的是"玻璃固化"这个表达。

③ 在翻译这个合成词的术语"low-activity waste，LAW"的时候，通过查询百度百科找到一段这样的文字："放射性废物是指含有放射性核素或被放射性核素污染，其放射性浓度或比活度超过国家规定限值的废弃物。放射性废物是指含有放射性核素或被放射性核素污染，其放射性浓度或活度大于审管机构确定的清洁解控水平，并且预期不再使用的物质。放射性废物根据放射性水平可分为：低放废物、中放废物、高放废物。低放射性废物是放射性废物中体积最大的一类，可占总体积的95%，其活度仅占总活度的0.05%。"这个词在《英汉核科学技术缩略语词典》给出的翻译本来是"低活废物"，但是从百度百科这段文字对放射性核废物根据反射性水平的分类来看，并没有低活废物，而是低放废物、中放废物、高放废物。看来尽管这个合成词里的"activity"本应该翻译为"活性"，比如"水活性，water activity"，"活性炭，active carbon"，但是本书最终还是采用了"低放射性废物"这个译文。

④ 这个句子里面出现了两处动词不定式被动语态的情况，它们的逻辑主语都是各自前面的那个"construction"短语，属于宾语补足语的成分。如果

把“被”字翻译出来，实属冗余，没有必要。如果省略“被”字，翻译成主动句，前面的逻辑主语“construction”结构只需通过增词，让它的词义具体化就可以在汉语中与动词构成主谓结构。第一个“construction of the PT Facility to be suspended”翻译为“预处理设施的建造工程暂停”，第二个“construction of the HLW Facility to be slowed”翻译为“高放射性废物处理设施的建造进度放缓”。

⑤ 这里的两个单词“combination”和“generation”属于英语中的名词化结构。科技文体属于一种非常正式庄重的文体。名词化结构不仅使科技英语表达精练、准确，而且使得整个语篇具有严肃性和权威性。它包括单纯名词化结构、复合名词化结构和动词或形容词派生的名词化结构。针对名词化结构的这些类别，选用合适的翻译转换策略，将之转译为名词词组、动宾结构、主谓结构和独立从句，从而实现准确、简洁和清晰表达科技信息的目的。把“combination”和它后面的名词成分处理成为动宾结构，翻译为“综合……”。对“hydrogen generation”则不调整词序，但是把名词化结构翻译为主谓结构“氢气生成”。

⑥ 一般来说，造成长句的原因有三方面：(1)修饰语过多；(2)并列成分多；(3)语言结构层次多。在分析长句时可以采用下面的方法：a. 找出全句的主语、谓语和宾语（主干/句），从整体上把握句子的结构。b. 找出句中所有的谓语结构、非谓语动词、介词短语和从句的引导词。c. 分析从句和短语的功能，例如，是否为主语从句、宾语从句、表语从句等，若是状语，它是表示时间、原因、结果，还是表示条件等。d. 分析词、短语和从句之间的相互关系，例如，定语从句所修饰的先行词是哪一个等。e. 注意插入语等其他成分。英语长句在翻译成汉语时，对原句的成分分析是很重要的。这个句子超过了 17 个单词，一般可看作是属于长句。主干是“complexity has led to issues”（复杂性导致了问题）。句子中的修饰成分有主语“complexity”后面的定语，宾语“issues”前面的形容词定语“difficult unresolved technical”和后面的“that...”定语从句。更为困难的是在“difficult”和“unresolved”之间还有个“to date”这个插入成分，以及宾语和定语从句之间的位于括号内的“facility”的同位语。面对这些困难，要根据英汉语言各自的特点在翻译时进行成分之间的对等处理或适当转换。本句的翻译就综合了句子成分对等和句子成分转换两种方法。比如，把“to date，unresolved”转换为句子的谓语，翻译为“至今尚未解决”，把

"complexity..."这个原来的句子主语转换为一个单独的分句，翻译为"核废物本身和废物处理与固化厂设施复杂"，而原来的谓语动词"has led to"转译为表示因果的连接词"以致于"。原来后置的定语从句也根据汉语的句子特点调整到被修饰名词的前面作定语。

翻译理论及应用解析：目的论

翻译目的论(Skopos Theory)是德国功能理论的核心理论。目的论的创始人汉斯·弗米尔(Hans Vermeer)在1978年《普通翻译理论框架》(*A Framework for a General Translation Theory*)首次提出目的论的三原则，即目的性原则(skopos principle)、连贯性原则(coherence principle)、忠实性原则(fidelity principle)。后来她的学生诺德(Christiane Nord)补充了"功能+忠诚"原则(the principle of funtion plus loyalty)。在这四个原则中，目的性原则是第一位的，即翻译的交际目的决定了采用何种翻译策略来实施翻译行为。连贯原则钳制忠实性原则，而两者都从属于目的性原则，即译者根据翻译目的必须首先考虑目的语读者的期望、接受程度和理解，使译文符合语内连贯的标准，而后再追求原文与译文的语际连贯。忠实性原则要求译者对原文作者和译文读者以及翻译委托人的意图和期望都要加以考虑。

可以看出，不管弗米尔的目的论三原则，还是后来诺德增加的忠诚原则，都不注重译文与原文是否对等，而是强调译者在翻译过程中发挥主观能动性。在以译文的预期目的为出发点的前提下，根据各种语境和文化因素，选择最佳的翻译策略，采用最合适的翻译方法，保证译文的连贯性和忠实性。还要对参与翻译活动的各方忠诚，以满足不同目的或需求的翻译委托人和译文读者的期望或预期。并且只要和原作者的交际目的不是相反或不一致，译文预期和实际要达到的目的或者功能可以不同于原文预期和实际达到的功能。也就是说，译者在目的性原则的指导下，根据连贯和忠实性的原则可以自主选择翻译手段和拥有更高的自由度，但也不可违反忠诚原则。

本章正文的英文原文选取自美国能源部2013年9月24日发布的《汉福德罐体核废料回收、处理、处置框架》(Hanford Tank Waste Retrieval, Treatment, and Disposition Framework)。原文是权威部门的专业人士撰写的专用科技文本，使用了专业性和程式化的语言，具有规范性、准确性和简洁性的特征。其译文的预期交际目的就是科技信息的交流、指导、警示和宣传的实用功能。这个目的论的目的性原则只从宏观角度指导科技翻译行为，从整体上明确翻译工作的指向，而译者要根据整个翻译过程参与者的需求，比如科技文本作者，科技翻译项目的发起者或者委托人，译文的接收者和使用者，包括科技译者自己的要求选取翻译策略以及具体的翻译方法，例如直译、意译、省译、改译和摘译等全译或者变译的方法。

例⑦原文“DOE took steps beginning in the 1980s to mitigate this risk by transferring all pumpable liquids from the older single-shell tanks to newer double-shell tanks.”译文为“从20世纪80年代开始，美国能源部采取了一些措施来降低这种风险，将所有可泵送液体从旧的单壳罐转移到新的双壳罐内。”这里的科技术语“single-shell, double-shell”在相关科技词典中没有对应的译文，但有“single-shell cylinder/单层缸”和“single-shell lining/单层衬砌”，以及“single-shell structure/单壳结构”和“single-shell dryer/单壳干燥器”。因此，这里虽然没有完美的功能对等的参考译文，但选择“单层”或“单壳”的译文都足以保证准确表达该专业词汇，最后译者自主选定的译文是“单壳罐/双壳罐”。此例的另外一个表达“pumpable liquids”中的“pumpable”在词典中给出的译文是“可用泵抽吸的”，但这样的说法不符合专业化的表达需求。为了保证译文的通畅连贯，又能简洁忠实地传达原文的意思，此处译为“可泵送液体”。此外，该例中的“beginning in the 1980s”为现在分词短语，其在句子中的位置割裂了谓语动词短语和它的动词不定式目的状语。使用割裂句式是科技英语句子中的特有结构，但在汉语中缺少对应的结构。如果按照英语的句式，译文读起来就会很费精力。因此要进行移化处理，即把它调整位置到最前面作句子的时间状语。

例⑧原文“Although there is no immediate health and safety risk posed by these leaks, addressing the long-term environmental concerns associated with the leaks requires a robust and sustainable strategy for waste retrieval, treatment, and disposal or long-term storage.”译文为“虽然这些泄漏不会带来直接的健康和安全风险，但是要解决与泄漏相关的长期环境问题，需要能源部针对废物回收、处理和处置或长期储存制定强有力的可持续战略。”原文全句是个复杂的长句，表达出一个复杂的科学论断。主从复合句是让步的逻辑关系，在汉语中采用“虽然……但是”表达。“There be”句型的内容必须转换为汉语的表达习惯，比如采用主宾互换的方式把介词“by”后面的内容译为上半句的主语，而原来的分词“posed”转换为该部分的谓语，其所修饰的名词短语成为译文中的宾语。同理,后半句中的两个名词后置定语"associated with..."和"for..."都移化到所修饰的名词前面作前置定语,译文表达结构更符合汉语受众的需求。

补充阅读:Power Generation and Energy Sources

The quality of power available through a power system not only depends on having an adequate amount of generation present, with a suitable reserve margin, but also the type of energy source used for the generation. It is important to model the physical control of the energy source when considering the effect the generator will have on system stability. In some cases, the design of a generating unit can be tailored to suit the stability characteristics required on the system.

1. Thermal Turbines

The energy source for thermal generation can be fuels (such as coal, gas, biomass, or any material that can be burnt to produce heat energy) or other heat sources (such as nuclear, geothermal, or solar-thermal energy). The energy produced is captured and converted by steam, gas or combination turbines.

For steam turbines, the stored energy in the source is used to heat water to generate high pressure, high temperature steam that is passed through a set of turbines. The turbine sets convert heat energy into rotational energy. The waste steam is then condensed and the water returned to the boiler where the cycle continues.

Gas turbines do not use an intermediate stage of water heating; instead they directly pass the hot, high-pressure combustion gasses through a series of turbine blades. Combined cycle turbines can increase efficiency by initially using a gas turbine stage, with the gas exhaust then used to heat water to feed a steam turbine.

Steam turbine thermal generation generally requires a significant amount of time to ramp up to full power output after an initial start, due to the time taken to raise the equipment to operating temperatures. After operating at constant output for some time, it can also take time to respond to a signal to ramp the power output up or down. This limits the ability of steam thermal generation to pick up extra load that may appear with little warning.

2. Photovoltaic

Photovoltaic (PV) generation involves converting energy from the light spectrum directly into electrical energy through the use of semi-conducting materials. Photovoltaic generation produces power in DC. This means the current does not alternate direction at 50 cycles a second, as in conventional AC generation. An inverter is used to connect to the power system.

The inverter can provide fast voltage control, even at night when the sun is unavailable. Photovoltaic generation has the advantage that the output is likely to be available during system peak loads as consumer demand is generally driven by temperature.

Although inverters could be capable of providing voltage control without output from the PV system they generally do not. There are issues with the fault ride through capability (ability to remain connected during faults) of a domestic connected PV system that could present issues as more PV systems are being connected to the distribution system. Also the power output from PV systems can be quite variable particularly on cloudy days. The power system, therefore, needs to be adequately designed to respond to uncontrollable variations in PV output.

3. Wind

A "wind farm" is typically made up of numerous turbine sets distributed over a moderate geographical area with each turbine producing a small amount of power, presently in the units order of 0.5-3 MW.

As wind is a renewable resource meaning that there is no fuel cost, there are initial capital cost and on-going maintenance costs. Wind generation has a variable output based on wind speed so cannot be scheduled using conventional techniques. This is a particular issue when the portion of wind generation is high, relative to the total generation pool. Other types of generation must be provided and are required to step in should the wind speed and resulting power output fall unexpectedly.

Traditionally wind farms have used induction generators; however, these will no longer meet the requirements of the Technical Rules unless they have significant reactive support.

Wind farms that use DFIG machines can provide voltage control although often within a limited range, and may not require additional reactive support to meet the requirements of the Technical Rules. In principle, inverter connected machines are able to provide full voltage control even when the wind is unavailable. This option is generally available at additional cost.

4. Cogeneration

Industrial or commercial plants with a requirement for steam/hot water now often include generating plant utilizing or producing steam to improve overall economics, as a "cogeneration", or "combined heat and power" (CHP) scheme.

The plant will typically have a connection to the public utility system; such generation is referred to as "embedded" generation. The generating plant may be capable of exporting surplus power, or simply reducing the importation of power from the utility. A cogeneration scheme directly coupled to the plant heating system has limited ability to vary its electrical output to meet power system needs without also varying the heat output to the rest of the plant. The scheme can have an auxiliary firing system fitted to allow the generator to have minimum electrical output even when the plant heat demand is high.

5. Water

Hydro generation converts potential energy into electrical energy using a water turbine connected by a shaft to the generating unit. The turbine can be designed for generation with differing levels of waterhead. The responsiveness of the generating unit to change power output varies significantly with the design of the turbine and penstock unit.

Compared with thermal generation, hydro generation is generally quicker to respond to small load fluctuations and can come online much faster.

6. Distributed Generation

"Distributed generation" involves the connection of small-scale power generation technologies to the distribution network. The power is usually generated close to where it is used. Often it is consumed on the network to which it is connected, without having to be increased in voltage for transmission over long distances.

Existing distribution networks have been developed in an environment where distributed generation was rare and small. Network design has taken advantage of simplifications that this allows (such as tapering the capacity towards the end of the network, low fault ratings, and the use of much simpler control and protection systems than are used on transmission networks). However, power generation using small-scale technology is increasing and connecting large amounts of generation to existing distribution networks is often problematic. Safety issues that arise include: ensuring adequate clearance of faults, and issues associated with the voltage profiles, conductor ratings and fault ratings. Many existing distribution networks require modification to cater for the new generation sources.

Chapter Five Electrical Construction and Maintenance Workers' Safety Manual

第五章 电力施工和维护人员的安全手册

原文

Before work begins in any confined area such as a manhole or vault, assess whether a hazard exists and determine if the area is a confined space. If the work area is a confined space, work must be performed under the appropriate confined space regulation. If the area does not meet the definition of a confined space but a hazard still exists, appropriate safety measures must be taken to address the hazard and to protect the worker. In some situations, the appropriate safety measure may be to treat the area as a confined space even if the definition does not apply.

译文

在进入例如下水道和地下室这样的密闭空间开始工作之前，要确定是否存在危险和施工区域是否为密闭空间。如果是密闭的，则工作必须遵循合适的密闭空间施工准则下进行。如果工作区域不满足密闭空间的定义并且存在危险，为了处理危险和保护工作人员，必须采取适当的安全措施。在某些情况下，即使对于密闭空间定义不是很明确的区域，也要采取适当的安全措施。

1 Authorized personnel in a ④Manhole, Vault, or ④Switchroom

1 检修孔、拱顶地下室和配电室允许的授权人员

Only authorized personnel are permitted access to a ④room or other ④enclosure containing exposed energized electri-

非授权人员不允许进入含有裸露的通电电气部件的密闭空间或者有围护墙的场地。

cal parts.

2 Trenches and Excavations

⑥ Before starting any excavation, ensure all utilities are located. Call the local utility companies.

3 Protection Against Cave-ins

There are three basic methods of protecting workers against trench cave-ins: • Sloping • trench boxes • shoring.

Most fatal cave-ins occur on small jobs of short duration such as service connections and excavations for drains and wells. Too often people think that these jobs are not hazardous enough to require safeguards against collapse.

• Sloping: One way to ensure that a trench will not collapse is to slope the walls. Where space and other requirements permit sloping, the angle of slope depends on soil conditions.

The type of soil determines the strength and stability of trench walls. ⑦ Identifying soil types requires knowledge, skill, and experience. Even hard soil may contain faults in seams or layers that make it unstable when excavated. The foreperson or supervisor must be knowledgeable about soil types found on a project and plan protection accordingly. This knowledge must include an awareness that soil types

2 沟槽和挖掘

在开始任何挖掘之前，要确定所有公共设施的位置。可以打电话给当地公用事业公司确认。

3 塌方防护

有三种抵御沟槽塌方的基本方法来保护工作人员：• 斜井 • 箱式沟槽支撑 • 横撑板

一些致命的塌方事故发生在细微工作的短暂瞬间，例如在为下水道和深井引入服务连接和挖掘排水的时候。通常人们认为这些工作没有危险到需要保护措施来避免坍塌的程度。

• 斜井：一种避免沟槽坍塌的方法是使沟槽的墙壁倾斜。空间和其他条件都允许倾斜时，坡度取决于土壤的条件。

土壤的类型决定了沟槽壁的强度和稳定性。鉴别土壤的类型需要知识、技能和经验。即使硬质土壤的矿层和岩层中也可能存在断层，这些断层会使沟槽在挖掘的时候变得不稳定。工头或者管理人员必须熟知工程中用到的土壤类型，并且要安排相应的保护措施。这些知识必须包含这样一种意识，土壤的

and conditions can change over very short distances. It is not unusual for soil to change completely within 50 meters, or for soil to become saturated with moisture over even smaller distances.

类型和状况会在很短的距离大不相同。土壤在 50 米之内完全改变或者在更小的距离内随着湿度变化而饱和是很正常的。

• Trench boxes : Trench boxes are not usually intended to shore up or otherwise support trench walls. They are meant to protect workers in case of a cave-in. ①If sloping is to be used above a trench box, the top portion of the cut should first be sloped according to the soil type (see above). Then the box should be lowered into the trench (Figure 13).

• 箱式沟槽支撑:箱式沟槽支撑通常不是为了加固或者支撑沟槽墙壁的。它们旨在出现塌方的情况下保护工作人员。如果斜井用在箱式沟槽支撑的上面,首先应该根据土壤的类型倾斜沟槽上层的切面。然后把沟槽支撑放入沟槽。

• Shoring: ② Shoring is a system which "shores" up or supports trench walls to prevent movement of soil, underground utilities, roadways, and foundations. Shoring should not be confused with trench boxes. A trench box provides worker safety but gives little or no support to trench walls or existing structures such as foundations and manholes holes. The two types of shoring most commonly used are timber and hydraulic. Both consist of posts, wales, struts, and sheathing.

• 横撑板:横撑板是指用来加固和支撑沟槽墙壁的系统,以防止土壤、地下公共设施、路基和地基移动。横撑板不能与箱式沟槽支撑混淆。箱式沟槽支撑给工作人员提供安全,但是对于沟槽墙壁或者像地基和下水道这样存在的建筑,它提供很少甚至于没有提供支撑。两种最常用的横撑板类型是木材的和液压的。两者都由柱材、横撑、撑杆和衬板组成。

4 Lockout and Tagging

4 锁闭和挂牌

② Lockout is the primary means of preventing the unplanned release of hazardous energy. For electrical workers, it often involves workers using a padlock

锁闭是防止意外释放危险能源的主要手段。对于电力工作者来说,它通常涉及使用挂锁的工作人员将开关保持在"关

to keep a switch in the "off" position. It may also be necessary to isolate the energy of moving parts, chemical reactions, falling counterweights, and other actions that can endanger lives. ④ Lockout is a physical way to ensure that the energy source is de-energized, deactivated, or otherwise inoperable. Lockout involves:

a. identifying all energy sources that may affect the work and work area

b. redirecting or stopping the energy from doing what it is normally intended to do

c. physically preventing the accidental reenergizing of the system, and

d. verifying zero energy.

⑦It is important to control all energy systems involved in the work. A piece of equipment may have an electrically-operated component but a hydraulic or pneumatic primary power source. Failure to control each energy system could jeopardize the safety of workers involved. In addition, gravity, momentum, and stored energy can present unexpected hazards.

Tags are an important part of a lockout. After attaching his or her personal lock, the worker attaches a tag to the lock. Tags are a means of communication. Tags are used to inform others that the device is locked out, who has locked it out, and why. ③ Tagged devices and

闭”位置。可能还需要隔离运动部件的能量、化学反应、重设备坍塌以及其他可能危及生命的行为。锁闭是确保能量源断电、停用或无法使用的物理方式。锁闭涉及：

a. 识别可能影响施工和施工区域内的所有能源。

b. 重新引导或阻止能源进行通常的预期行为。

c. 物理防止系统的意外重新启用。

d. 验证零功率。

对工作中涉及的所有能源系统控制，这非常重要，因为设备可能除了有电动操作部件，还有液压或气动主动力源。不能控制每种能源可能会危及工作人员的安全。另外，重力、动量和储存能量可能会带来意想不到的危险。

挂牌是锁闭的重要组成部分。在附加个人锁闭后，工作人员将挂牌粘贴到设备上。挂牌是一种沟通方式。挂牌用于通知其他人该设备被锁闭，谁锁闭，以及为什么。如果没有挂牌上标明专业人员的授权，贴挂牌

systems must not be re-energized without the authority of those named on the tag.

的设备和系统不可被重新通电。

⑤ It is absolutely essential for an employer to have an effective lockout and tag policy in place and to ensure it is diligently practiced without exception.

绝对有必要让雇主制定有效的锁闭和挂牌政策并确保其毫无例外地被积极遵照执行。

5 Forms of Energy

5 能量形式

When most people think of uncontrolled hazardous energy, they think of electricity. But electricians overseeing a lockout procedure need to consider a variety of energy sources. Here are the main types of energy.

大多数人想到不受控制的危险能源时，会想到电力。但监督锁闭程序的电工需要考虑各种能源。这里是主要的能源类型。

- Electrical (electrical panels, generators, lighting systems, storage batteries, etc.)

- 电能(电气面板、发电机、照明系统、电池等)。

- Mechanical—the energy of moving parts (flywheels, blades, fans, conveyor belts, etc.)

- 机械能——运动部件的能量(飞轮、叶片、风扇、传送带等)。

- Potential—stored energy that can be released during work. Examples of systems having potential energy include suspended loads, compressed air, electrical capacitors, accumulated bulk goods, coiled springs, chemical reactions, changing states (solid— liquid—gas), etc.

- 电势——工作期间可能释放的潜在储存能量。具有电势的系统包括悬浮负载、压缩空气、电容器、螺旋弹簧、化学反应、状态变化(固体—液体—气体)等。

- Hydraulic (presses, rams, cylinders, cranes, forklifts, etc.)

- 液压能(压力机、油缸、气缸、起重机、叉车等)。

- Pneumatic (lines, compression tanks, etc.)

- 气动能(管线、压缩罐等)。

- Thermal (steam, hot water, fire,

- 热能(蒸汽、热水、火灾

etc.)

• Chemical (flammable materials, corrosive substances, vapours, etc.)

Some equipment may involve more than one type of energy, and pose unexpected hazards. ④ For example, a machine may have an electrically operated component but a hydraulic or pneumatic primary power source, or it may become activated on a timed schedule. With some equipment, gravity and momentum can present unexpected hazards.

③ A de-energized electrical system must be discharged by short circuit and phase to ground. A temporary ground cable must be attached to the system and remain in place until work is completed.

Switches, power sources, controls, interlocks, pneumatics, hydraulics, computer-controlled sources, gravity-operated sources—all of these must be locked out by each worker involved and appropriately tagged. Recognize and control the energy associated with them.

等)。

• 化学能(易燃材料、腐蚀性物质、蒸汽等)。

某些设备可能涉及多种类型的能源,并造成意想不到的危险。例如,一台机器可能有一个电动操作部件,也有一个液压或气动主动力源,或者它可能会按照时间表启动。使用某些设备时,重力和动量会带来意想不到的危险。

必须为断电的电气系统通过短路和相对接地放电。必须将临时地线连接到系统并保持到位,直到施工完成。

开关、动力源、控制装置、连锁装置、气动装置、液压装置、计算机控制源、重力操作源——所有这些都必须由每个涉及的工作人员闭锁并适当挂牌,以识别并控制与之相关的能量。

《术语表》

electrical construction	电力施工
confined space	密闭空间
manhole	人孔,检修孔
vault	拱顶地下室
switchroom	配电室
sloping	斜井

trench box	箱式沟槽支撑
shoring	横撑板
lockout	锁闭
electrically-operated component	电动操作部件

翻译难点讲评

① 对于大量专业化了的普通词汇，直译往往不能满足通顺表达的要求，甚至会产生歧义。这样，在翻译时就需要做出相应调整，或改变词语词类，或选择词汇的引申义，或进行适当的变译。有时，也需要根据目的语的文化背景和语言特点，进行增译或省译。这里的"Cut"显然不能按照其普通词义"伤口"进行翻译，它的专业词义也比较多，"横切矿脉［矿业］；截除［计］；锯解［木］；每匹经纱长度［纺］；切［科技］切割［机］；切换［电影电视］；切削［机］"。这里所讲的内容是开挖沟槽，在放入箱式沟槽支撑之前，如果上面是斜井，那么沟槽的上层部分的"cut"应该根据应用领域和上下文翻译成"切面"。

② 科技英语文本中名词化结构使用率高，在传递信息时，能够既保证行文的简洁度，又客观确切地表达内容。然而相对于英语多静态的特点，汉语是一门动态语言（连淑能，2010），所以在翻译名词化结构时，需要进行必要的转化和调整，以符合汉语表达习惯。笔者在文本翻译时最常遇到名词化结构形式为"行为名词 + 介词 + 名词"，如"release of hazardous energy"。这一抽象名词来强调存在的事实而非行为。在翻译时将其还原为词"释放"，这样原来的名词化结构转化为动宾短语，译文读起来也更加自然。有时，也可以将这类名词化结构转化为主谓结构，如"movement of soil, underground utilities, roadways, and foundations"在翻译成汉语时，可以将其转化为主谓结构，即翻译为"防止土壤、地下公共设施、路基和地基移动"。

③ 在科技文体中被动语态的使用频率高于普通文体，这与其内容的严肃性、客观性是分不开的。在翻译时，为维持文体原有风格，最好也译成"无主语结构或被动意义的句子"（穆雷，2008）。另外，被动语态翻译成汉语的主动语态同样较为普遍，因此在翻译时可灵活采取保留原文中的被动语态或者转换为汉语主动语态或者无主句等。例如："Tagged devices and systems must not be re-energized without the authority of those named on the tag. 如果没有挂牌上标明专业人员的授权，贴挂牌的设备和系统不可被重新通电。（翻译为汉语被动句）”和“A de-energized electrical system must be discharged by short circuit and phase to ground. 必须为断电的电气系统通过短路和相对地放电。（翻译为汉语无主句）”。

④ 文本词汇层面最鲜明的特征莫过于其丰富的专业词汇，充分体现了科技文体的科学性、技术性和专业性。并且电力专业英语覆盖面广，文本涉及机械、仪表、电气、电子、计算机、物理、土建等专业，因此在翻译时需要注意辨析词汇在语境下的真正含义，避免出现外行话而影响到译文的表达效果。有的专业词汇词义清晰且较为固定，如上文提到的电力专业特有词汇，在翻译时可以采用直译的策略，使用行业内约定俗成的表达，如：“lockout，锁闭；manhole，人孔、检修孔；switchroom，配电室”。部分电力与其他专业共用的词汇，在翻译时必须根据文本内容选择相应的义项，如“room”一般用作名词为“房间；空间；余地；机会”，在电力英语中则根据上下文赋予其词义，这里指的是“密闭空间”。又如名词“enclosure”在一般用语中为“附件；围墙；围场”，在电力英语中为“外壳，围墙，附件，外厢，围绕，围绕物，围护墙，四壁，套，罩，护栏，包围，封入，防护方式，包封，封管，封闭，包壳，包封设备，封入物”，而在此处上下文中则为四周有墙壁环绕的空间，译为“围护墙”。这些都可在电力专业词典中查证得出。术语词典的使用并不意味着其可以代替译者的思考，比如词组“a hydraulic or pneumatic primary power source”的翻译。术语词典中包含了“hydraulic”和“pneumatic”两个词，词义分别为“水力的，液压的，油压的，水压的”和“气动的”，翻译到此处时可跳过查询工具书的过程，但仍需根据上下文选用合适的词汇进行表达。而术语词典中“primary”的词义为“初级线圈，一次线圈；主要的，原始的，原，初，基本的；初级，初级的；原色；最初的，原有的；一次的，一回路的；初次的”，另有“primary battery/cell”意为“原电池”，但综合来看，这些释义在此处均不适用。考虑到有“primary chamber，主燃烧室；primary（re）circulation pump，主[再]循环泵”，故将“primary power source”翻译为“主电源”或“主动力源”比较合适。

⑤ 英语习惯上常常以引导词 it 作形式主语，而把真正作主语的动名词、不定式或主语从句放在谓语之后。此时，it 本身无词义，汉译时无须译出。这种句型是英语中特有的一种句型，也是科技英语中常见的一种句型。这种句型的译法多种多样，概括一下有两种译法：顺译法和倒译法。顺译法

就是基本按照原文句子的语序，将含有 it 作形式主语的英语句子译为汉语的无主句。倒译法就是将位于句尾的真正的主语(包括主语从句、不定式短语、动名词)先译，并在译文中作主语。本句采用的第一种翻译方法，把 it 形式主语句子翻译为汉语的无主句，整个句子采用顺译，基本没有改变原句的语序："It is absolutely essential for an employer to have an effective lockout and tag policy in place and to ensure it is diligently practiced without exception. 绝对有必要让雇主制定有效的锁闭和挂牌政策并确保其毫无例外地被积极遵照执行。"

翻译理论及应用讲评：动态/功能对等理论

在 20 世纪 60 年代，尤金·奈达(Eugene Nida)在 *Toward a Science of Translating* 及 *The Theory and Practice of Translation* 中提出了"动态对等"(Dynamic Equivalence)的概念。70 年代，奈达直接用"功能对等"(Functional Equivalence)替代了"动态对等"。奈达给动态对等下的定义："在译语中用最贴切而又最自然的对等语，再现原语的信息，首先是信息，其次是风格。"这里的信息是广义的，不单指语义信息，还包括文体信息、情景信息和心理信息在内的所有原有信息。由此看，奈达的动态对等观念实质上是某种程度上的归化翻译。奈达给功能对等的定义："译文读者应能以原文读者基本相同的方式理解和欣赏译文，此为最高限度对等；译文读者至少应能通过对译文的理解而想象得到原文读者是如何理解和欣赏原文的，此为最低限度的对等。"由此看，功能对等除了突出文本内容信息的对等，更强调语用和语境等交际功能的对等。可见，功能对等非常重视译语读者的接受程度和读者反应。奈达在 *From One Language to Another*：*Functional Equivalence in Bible Translation* 一书中提出的功能对等实质为动态对等的修正版，二者并无实质区别。功能翻译要求内容对等，形式也要对等。如果翻译时，原文和译文不可避免地发生冲突，在动态/功能对等中形式要让位于内容，但这并不是说形式本身无价值。对等原则，即要在词汇、句法、修辞、文化、语义和文体风格等各个层次上，通过改变或保留原文形式，保存原文内容，在译语中采用最通畅和简洁的语言表达出来。

科技英语中含有大量来自普通英语的词，其基本词义经过了扩大或缩小或转移，所以会有一词多义、一词多领域使用的情况。而汉语给某一专业概念命名的基本趋势是避免使用常用词语,尽量做到一词一义。因此,科技英语的一词语多义、多专业化倾向在科技汉语中是不多见的。也就是说,英语术语的一个词的多种意义在汉语中往往用各专业不同的词语来表达。所以,翻译科技词语要表达的是其某一专业化、具体化的符合语境的词义,而不是通用词义或者其他专业词义。按照奈达的动态/功能对等理论,译者要寻找的是语义上的对等语,而不是形式上的对应语。如例⑥原文"Before starting any excavation, ensure all utilities are located. Call the local utility companies."译文为"在开始任何挖掘之前,要确定所有公共设施的位置。可以打电话给当地公用事业公司确认。"这句中的"utility"一词出现在两个位置,该词的词义有"公共/用设施;电力公司,电站;效用;公用事业;实用;利用"等。本句中译者分别选用的是"公共设施"和"公用事业"。虽然本文是电力专业领域的文本,该词在电力领域有对应的专业领域词义"电力公司"和"电站"。但是根据本句的语义,电力施工本来就是电力公司员工的工作,绝无再向电力公司打电话的必要。另外,上下文语境显示此处所说的电力施工不是电站建设之类的大规模施工。

英汉语存在语言表达习惯上的差异,表达同样的语义会采用不同的语言结构。奈达所说的功能对等有程度高低之分,科技翻译追求的是最贴近的对等,即与原文意义最接近的符合译语表达规范的译文。如例⑦原文"It is important to control all energy systems involved in the work. A piece of equipment may have an electrically-operated component but a hydraulic or pneumatic primary power source."。译文为"对工作中涉及的所有能源系统控制,这非常重要,因为设备可能除了有电动操作部件,还有液压或气动主动力源。"此例原文中的"but"所表达的逻辑关系在英语中很常见,前后两个句子之间没有使用连词。如果受限于此结构形式,在译语中照搬或者简单复制原文中的结构,达到"形合",就会产生带翻译腔的译文。在该例的译文中,首先进行了合译,对隐含的逻辑关系通过增译"因为"来处理;其次对于表达转折的"but",根据前后语义,转换成了表达递进关系的"除了……还……"。把隐含的逻辑关系显化处理,把原文一种的连接关系译为译文中的另一种连接关系,目的是功能对等,结果也更符合汉语表达习惯,更易于读者理解。而如果一味追求形式上的对应,效果只能适得其反。另外,此例也显示出,实现语句结构的功能对等,有必要对原语的语句结构进行调整或转换,而变动的程度和范围应依照译语的表达规范来确定。

补充阅读:Power Quality

The power system needs to run in a stable state and be able to deliver power to consumers at an acceptable quality. All connected equipment must operate in the desired manner.

The correct operation of a network requires the voltage to stay within an operational range, which all electrical equipment is designed to handle.

Broadly, the essential qualities are:

• sinusoidal waveform, with a low level of disturbances (from motor starting, harmonics etc.)

• a constant power system frequency of 50 Hz.

At times of heavy loading on the power system, the current flowing through any piece of equipment should not be above its design rated level. During a fault all equipment must be able to handle the current.

The impedance of generators, transformers, and transmission circuits should be low enough to allow power to flow to where it is required without excessive voltage drop from generation to load. The impedance should be sufficient to ensure that the fault current on any one piece of equipment will not be excessive.

A new generator has the responsibility to ensure that they can output power across their full output range, while meeting the power quality requirements of the Technical Rules.

1. Voltage

The voltage at which the grid is designed to operate is referred to as "nominal voltage". Nominal voltage can also be referred to as 1 per unit (pu) or 100 per cent (of nominal voltage). When a point at the grid is operating at 1.05 pu or 105 per cent, this means that the voltage measured at that point is 5 per cent higher than the nominal voltage. The voltage varies on a power system depending on the size and location of loading, and the route the power has to travel from generation to load.

Normally, if the voltage at any point on a network rises above 110 percent or drops below 90 percent, the quality of power delivered is unacceptable as it may cause damage to connected equipment.

Closer to the power consumer, the range of acceptable voltage variation during normal operation decreases to plus or minus 6 per cent.

Transformers play a vital role in voltage regulation by changing their voltage conversion ratio (called "tap changing") depending on loading conditions of the system to keep the system within voltage limits. However, transformers take time to change taps, and are not able to smooth out surges and other short-term voltage variation.

The quality of power transfer on a network may also be poor if the voltage varies erratically. The most common problems are:

• Flicker. Large loads connected to the network that vary or switch regularly can cause small voltage variations on the surrounding network. These cause brightness of incandescent lights to change, which is irritating to users.

• Dips. Starting a large motor on a weak network, usually the distribution network, can cause a large voltage dip lasting up to ten seconds. A dip in voltage can cause other loads connected nearby to switch off. In severe cases, damage may be caused.

• Harmonics. The connection of large variable speed drives to a network introduces power noise onto the surrounding network. This is called "harmonics". The current flow caused by harmonics can cause other electrical devices to overheat and malfunction. The prolonged presence of harmonics on a system will shorten the life of sensitive equipment.

• Unbalance. Voltage unbalance will cause heating of three phase machines and reduction in motor output torque which leads to increased machine losses and reduced efficiency.

Some reasons for voltage unbalance are: unequal loading on each of three phases of the power system; incomplete transposition of long transmission lines; and railway traction loads.

2. Frequency

The frequency of the power system directly relates to the rotational speed of machines connected to the power system. The frequency needs to be maintained to avoid damage to motors. If the frequency drops, normally due to load increase, generation must be increased to bring the frequency back to normal levels and vice versa. Generation needs to be able to run over the full range of allowable network frequency, as the unscheduled trip of a generator can decrease system frequency.

3. Current

The current flowing through a given circuit in the network should not exceed the rating of connected equipment. Currents above the rated current can cause equipment to suffer excessive mechanical stress and overheat. This may result in permanent damage to equipment.

A fault at some point along a circuit can cause large currents to flow in the network to the fault point. The equipment surrounding the fault needs to be capable of handling large currents without overheating until protection can isolate the supply of power to the fault.

4. Impedance

The impedance of an electrical circuit needs to be low enough so that the generator can supply the maximum desired export power without causing the voltage to drop too low at either end.

Occasionally, when new generation is connected to strong points on the network, usually near the existing generators, additional impedance may be required to prevent the fault contribution of the generator from exceeding the fault ratings of nearby network equipment.

5. Power

Occasionally, limits are placed on generator power output. This may occur during the outage of an electrical circuit, which helps transfer the generator power output to the load centres, or when voltages on the system reach the limits of allowable range. Typical operating constraints are normally discovered before a new generator is installed by simulating a range of operating scenarios with proposed generation. If limits are considered too restrictive other electrical equipment can be installed to increase the range of operation.

6. Complex Power or Apparent Power

"Complex power" is the combination of the power required to enable power transfer across a power system (the reactive, or imaginary part) and the useable power being transferred from generation to load (the real part). The real and reactive power parts are discussed separately as they behave differently on the power system.

6.1 Reactive Power

"Reactive (or imaginary) power (kVar) provides the force that enables power transfer across the power system. It supports the voltage necessary to drive real power through the electrical circuits to the areas of load. Transformers and many motors absorb reactive power, and use it to establish a magnetic field, allowing energy to flow through the magnetic coupling. Generators and capacitors produce reactive power, and their output is balanced to meet total reactive demand on the power system. Generators can provide a variable level of reactive power by changing the excitation of the rotor, and can provide continuous voltage control with the help of voltage monitoring control circuits. Capacitors are fixed devices that can switch in blocks to provide a set level of reactive power and are used to boost the voltage.

Reactive power support is required to stabilize the voltage. If there is insufficient reactive power the voltage will reduce. If the shortfall is great enough, it will lead to voltage collapse. If there is too much reactive power, then there are issues with over voltages. This is a particular problem with long lightly loaded lines.

6.2 Real Power

"Real power" (kW) is the power available on a power system to be converted by loads into other forms of energy. The real power output of generation must be balanced by the power consumed by loads and losses. If the generation is greater than the energy used, then the frequency will increase. If the generation is less than the energy used, then the frequency will fall. To maintain a constant power system frequency, the generation must be exactly matched to the loads and losses.

Chapter Six Uninterruptible Power Supplies

第六章 不间断电源

原文

① Power ② distortions such as ① power interruptions, voltage sags and swells, voltage spikes, and voltage harmonics can cause severe impacts on sensitive loads in the electrical systems. Uninterruptible power supply (UPS) systems are used to provide uninterrupted, reliable, and high quality ① power for these sensitive loads. Applications of UPS systems include medical facilities, life supporting systems, data storage and computer systems, emergency equipment, telecommunications, industrial processing, and on-line management systems. The UPS systems are especially required in places where power outages and fluctuations occur frequently. A UPS provides a backup ①power circuit to supply vital systems when a power outage occurs. In situations where short time ① power fluctuations or disturbed voltage occur, a UPS provides constant ① power to keep the important loads running. During extended ①power failures, a UPS provides backup power to keep the systems running long

译文

功率畸变，例如电源中断、电压骤降与涌浪、电压尖峰和电压谐波，会给电气系统中的敏感负载造成严重的影响。不间断电源(UPS)系统用于给这些敏感负载提供不间断的、可靠的、高品质的电源。不间断电源系统的应用包括医疗设备、生命支持系统、数据存储和计算机系统、急救设备、电信、工业加工和在线管理系统。停电和电压波动频繁发生的地方尤其需要不间断电源系统。当发生停电时，不间断电源系统提供了备用电源线路来支持重要系统用电。当发生短时间电压波动或电压干扰情况时，不间断电源系统提供了恒定功率以保持重要负载运行。在长时间电源故障期间，不间断电源系统提供备用电源来维持系统运行足够长的时间以便平稳断电。

enough so that they can be gracefully powered down.

Most of the UPS systems also suppress ③ line transients and harmonic disturbances. Generally, an ideal UPS should be able to simultaneously deliver uninterrupted power and provide the necessary power conditioning for the particular ① power application. Therefore, an ideal UPS should have the following features: regulated sinusoidal output voltage with low total harmonic ② distortion (THD) independent from the changes in the input voltage or in the load, on-line operation that means zero transition time from normal to back-up mode and vice versa, low THD sinusoidal input current and ① unity power factor, high reliability, high efficiency, low EMI and acoustic noise, electric isolation, low maintenance, low cost, weight, and size. ⑤ Obviously, there is not a single configuration that can provide all of these features. Different configurations of UPS systems emphasize on some of the features mentioned above.

大部分的不间断电源系统也能够抑制线路瞬变和谐波扰动。一般来说,理想的不间断电源系统可以持续地提供连续的电力并给特定的电力应用提供必要的功率调节。因此,理想的不间断电源系统应当具备以下特性:稳定的正弦输出电压,不受输入电压或是负载变化影响,全谐波畸变(THD)较低。联机操作意味着从正常模式到备份模式的转换时间为零,反之亦然。THD正弦输入电流低和单位功率因数低、可靠性高、效率高、电磁干扰(electromagnetic interference,EMI)低和噪声低、电子隔离、维护低、费用低、重量小、尺寸小。显然,一个单一配置的不间断电源系统无法具有上述所有特性,侧重于对上述某些特性的差异化配置。

Standby UPS

This configuration of UPS system is also known as "off-line UPS" or "line-preferred UPS". It consists of an AC/DC converter, a battery bank, a DC/AC inverter, and a static switch. A passive low pass

备用不间断电源系统

不间断电源系统的配置也被称为"离线不间断电源"或"线路首选的不间断电源"。它包括AC/DC交流—直流变换器,一个电池组,一个DC/AC直流—

filter may also be used at the output of the UPS or inverter to remove the switching frequency from the output voltage. The static switch is on during the normal mode of operation. ④ Therefore, load is supplied from the AC line directly without any power conditioning. At the same time, the AC/DC rectifier charges the battery set. This converter is rated at a much lower power rating than the power demand of the load. When a power outage occurs or the primary power is out of a given preset tolerance, the static switch is opened and the ① DC/AC inverter provides power to the load from the battery set for the duration of the preset backup time or till the AC line is back again. This inverter is rated at 100% of the load power demand. It is connected in parallel to the load and stays standby during the normal mode of operation. The transition time from the AC line to DC/AC inverter is usually about one quarter of the line cycle, which is enough for most of the applications such as personal computers. The main advantages of this topology are simple design, low cost, and small size. On the other hand, lack of real isolation of the load from the AC line, no output voltage regulation, long switching time, poor performance with non-linear loads, and no line conditioning are the main disadvantages of this configuration.

交流逆变器和一个静态开关。无源低通滤波器也用于不间断电源系统的输出或逆变器来从输出电压中移除开关频率。静态开关在正常工作模式处于ON开启状态。因此，直接从交流线路为负载供电，不需要任何电源调节。同时，AC/DC交流—直流整流器对电池组进行充电。该变换器的额定功率远低于负载所需功率。当断电发生或主电源不在给定的预设容许偏差范围内时，静态开关断开，DC/AC逆变器在预设备用时间内从电池组向负载供电，直至交流线来电。逆变器的额定功率100%按照负载的功率需求。它与负载并联，在正常工作模式下保持备用。它从交流线路到DC/AC变换器的转换时间通常约为线路周期的1/4，这对于大多数个人计算机等应用设备来说是足够的。这个拓扑的主要的优点是设计简单、费用低、体积小。另一方面，负载与交流线之间缺乏真正的隔离、没有输出电压调节、开断时间长、非线性负载的性能不佳以及缺乏线路调节是这个配置最主要的缺点。

Different configurations of AC/DC rectifiers such as linear or switching may be used in this system. To reduce the cost, a simple diode-bridge rectifier with a capacitor at the front end is used. A full-bridge or half-bridge full controlled converter is also used to charge the battery bank. The full controlled topologies can provide ① power factor correction (PFC) to meet the corresponding standards. ④ To optimize the charging process, the charging cycle is divided into "constant current" and "constant voltage" modes. In the constant current mode, the converter injects a constant current into the battery till the battery is charged up to about 95% of its capacity. After this mode, the constant voltage mode starts that applies a constant voltage on the battery. In this mode, the input current of the battery declines exponentially until it is fully charged.

在系统中可以使用不同配置的AC/DC整流器(如线性型或开关型)。为了降低成本,使用了一种前端有电容器的简单二极管桥式整流器。全桥或半桥全控变流器也用来对电池组充电。全控的电路拓扑结构可以提供功率因数校正(PFC)以满足相应的标准。为了优化充电过程,充电过程分为“恒流”和“恒压”两种模式。在恒流模式下,变流器向电池中注入恒定电流,直到电池充电到其容量的95%左右。接下来恒压模式启动,即在电池上施加恒定电压。在恒压模式下,电池的输入电流呈指数型下降,直到电池充满电。

The purpose of the DC/AC inverter is to provide high quality AC power to the load when the static switch is opened. A full- or half-bridge topology is used for this inverter. In some topologies of standby UPS systems, an isolating transformer is used at the output stage of the UPS. This topology is called ferroresonant standby UPS system. The transformer also acts as a low pass filter that cancels out switching frequency from the output voltage of the

DC/AC逆变器的作用是当静态开关打开时向负载提供高质量的交流电。DC/AC逆变器是通过全桥或半桥的拓扑电路结构来完成的。在一些备用电源拓扑结构中,UPS的输出级安装了隔离变压器,这种拓扑电路称为铁磁谐振备用不间断电源系统。隔离变压器还起到了低通滤波器的作用,滤除由DC/AC逆变器产生的输出电压开关频

DC/AC inverter. On the other hand, the transformer stores electromagnetic energy in the core and acts as a buffer when a power outage occurs. ①For a short time, the transformer provides power to the load and protects sensitive equipment from being affected during the transfer time from the input AC to the UPS. Since the transformer is bulky and expensive, this configuration is more appropriate for high ① power applications.

率。另一方面,变压器在铁芯中储存电磁能量,在停电时充当缓冲器。在短时间内变压器向负载提供电能并且保护敏感设备,避免其在从交流电切换至UPS电源输入的过程中受到影响。由于变压器体积庞大且价格昂贵,这种配置更适合于高功率应用设备中。

《术语表》

power distortion	功率畸变
harmonic distortion	谐波畸变
UPS	不间断电源
EMI	电磁干扰
sinusoidal output voltage	正弦输出电压
AC/DC converter	交流/直流变换器
DC/AC inverter	直流/交流逆变器
static switch	静态开关
AC/DC rectifier	交流/直流整流器
ferroresonant standby UPS	铁磁谐振备用不间断电源
power factor correction	功率因数校正
isolating transformer	隔离变压器
constant current	恒流
low pass filter	低通滤波器
preset tolerance	预设容许偏差
constant voltage	恒压

翻译难点讲评

①"Power"在几乎所有电气类文本或者与电气相关的文本中出现的频率都

是比较高的，本文的前两段“power”就出现了10次以上，而“power”本身具有多重意义，词典中给出的释义有：“动力，能力，功率，（电）能，电源，幂，电（力），拖动，权”。“电源，电（力），（电）能，功率”这几个义项也是在所有电气类文本中最常见的。当“power”译为“（电）能”之意的时候，其意义相对来说是比较好确定的，译者只需要有一些常识储备就可以推敲出其词义，当然，个别情况下还是需要译者有一定的相关知识的储备，比如“the transformer provides power to the load”，只要了解“能量”的基本特性，就是可以说“变压器向负载提供电能”，然后就可以比较容易地去处理原文相关内容了。“Power”译为电源的情况在本文中主要是提到不间断电源 UPS 时，或者提到“power failure 电源故障”，或者提到“power outage 电源断电”，或者“power interruption 电源中断”，或者“backup power circuit 备用电源线路”这几种情况。而对于“power”的另外两个词义“电（力）”和“功率”，其词义的选择相对有些难度，特别是涉及相应的科技原理解释的时候，对于没有专业知识背景的译者来说很容易混淆，所以，如何正确处理易混淆的词义选择，对译者正确翻译原文起到举足轻重的作用。比如一种情况是在专业术语中出现，“The full controlled topologies can provide power factor correction（PFC）to meet the corresponding standards.”这里的“power”的词义相对比较好确定，因为它出现在一个专业术语即“power factor correction (PFC)”中，这样我们就可以通过查阅相关专业词典来确定其词义为“功率因数校正”，而不用做过多考证。“power”译为“功率”的情况，首先，根据句意分析，如果是专业术语的话可以直接在相关词典查出，其次，要根据相关的科学原理来深层分析，判断好各个参量之间的联系，比如可以根据功率的公式确定其意，前文提到过，无论是瞬时功率，抑或是平均功率都涉及两个参量——电流和电压，因此，如果上下文中提到了电流和电压的话，那么就可以确定为“功率”之意。比如“power distortion 畸变功率”和“constant power 恒定功率”。另外，也可按照电气常识来判断，通常用电设备功耗等级都是以功率大小来判定的，所以，这时都译为“功率”之意，比如“high power application 高功率应用设备”。最后，“power”的词义“电（力）”一般情况下都可以通过常识或者查阅相关词典来确定，但有的时候也要通过仔细分析科学原理的实质来确定其意义，也可以利用权威的学术网站，通过引用频率的比对来确定其词义。比如“DC/AC inverter provides power to the load from the battery set for the duration of the preset backup time or till the AC line is back again. DC/AC 逆变器在预设备用时间内从电池组向负载供电，直至交流线来电。"这里只需在搜索引擎中查找逆变器的工作原理就可以得到相关信息，结合知网翻译助手可核实"供电或者提供电力"的使用频率来确定译文。

② 通用科技词汇不同于上文提到的普通词汇和专业词汇。通用科技词汇既不像普通词汇那样在日常生活中使用频率那么高，也不像专业词汇应用范围那么窄，只应用于某一专业或者行业当中，所谓的"通用科技词汇指不同专业都要经常使用的词汇，但在不同的专业内却有不同的含义"（杨丰宁，2006）。"Distortion" 作为一个通用科技词汇出现的频率不是那么高，虽然词典上给出的词义和使用范围都很明确"[物][计]畸变；[电子]失真"，但由于搭配形式不同，词义的选择往往不像表面上看似那么容易。以下是与"Distortion"搭配时其被翻译成"畸变"和"失真"的例子："intermodulation distortion 互调失真；phase distortion 相位失真；harmonic distortion 谐波畸变；attenuation distortion 振幅失真，衰减畸变；distortion factor 失真因数，失真度；lattice distortion 点阵畸变；image distortion 图像失真；systematic distortion 基本畸变，规则性失真；amplitude distortion 振幅失真，波幅畸变；angular distortion 角失真，角度畸变；elastic distortion 弹性畸变；distortion meter 失真度测试仪，失真测；pattern distortion 图像失真；aperture distortion 小孔畸变，孔径畸变；peak distortion 峰值失真；barrel distortion 桶形失真，桶形畸变；waveform distortion 波形失真，波形畸变；line distortion 线路失真，行失真；linear distortion 线性失真；field distortion 场畸变，磁通分布畸变"。那么，"power distortion"是翻译为"功率畸变"，还是"功率失真"呢？在知网上搜到一篇文章：曹玉泉，白丽丽，颜伟中．用功率四边形分析非正弦交流电路的功率因数[J]．西北农林科技大学学报（自然科学版），2006，34(2)：153-156．，该文章中使用的"distortion power"是一种物理量，文章对于非正弦交流电路中存在的畸变功率如何影响到了视在功率、有功功率、无功功率之间的关系等物理量的表达式做了分析。因此，在物理领域中还是要使用的是"畸变功率"的表达。该例子说明，通用科技词汇因为在不同的专业中有不同的意义，有的时候在同一个专业文本中还有可能出现不同的释义，所以在翻译通用科技词汇时，首先，确定所翻译的文本所属专业类型，这样会缩小选择的范围，其

次，一定要仔细研究此类词汇出现的最原始的范畴，不能只从表面上判断其出现的专业范围就武断地确定其释义。

③ “Line transients and harmonic disturbances”这两个合成词术语组成的两个部分都有来自普通英语词汇的单词。由普通词汇和其他词汇组成专业术语的这种创造新的行业术语的方式已经十分普遍，这样组成的合成词专业术语比比皆是，在各行各业中都有许多实例。每个专业在刚诞生之初，都有一些属于自己专业的术语，慢慢地随着对本专业的深入研究和发展，有越来越多新的专业术语出现，而由普通词汇构成专业术语的方式大大地扩充了专业术语的数量，这种专业术语的发展与扩充有助于促进本行业的发展与壮大。单词“transient”的专业词义可以在词典中查到“瞬态，瞬变，暂态”。单词“disturbance”的专业词义可以在词典中查到“骚扰，干扰，扰动”。通过知网工具书查询，可以找到“line transient”在工具书“白英彩. 英汉计算机技术大辞典：新世纪版[M]. 上海交通大学出版社，2001.”中出现，对该词条的翻译为“线路瞬变”。通过知网工具书查询，可以找到“harmonic disturbance”在工具书“刘斌蓉. 英汉电网技术词典[M]. 四川科学技术出版社，2012.”中出现，对该词条的翻译为“谐波扰动”。由此可以看出，对普通词汇组成专业术语的翻译，首先，根据词组出现的具体语境，以及词汇本身所属的专业领域范围进行翻译。也可以利用权威的学术网站、网络工具书和电子词典等科技手段来确定词义。

④ 电气文本中被动句的翻译不能完全按照英文原文本的形式照搬全抄，而是要在准确传达原文信息的前提下，按照汉语的使用习惯进行翻译。对于英文的被动句，通常都会处理成汉语的三种形式，即译为汉语的主动句，用汉语的主动形式表达英语的被动意义，比如“To optimize the charging process, the charging cycle is divided into ‘constant current’ and ‘constant voltage’ modes. In the constant current mode, the converter injects a constant current into the battery till the battery is charged up to about 95% of its capacity. 为了优化充电过程，充电过程分为‘恒流’和‘恒压’两种模式。在恒流模式下，变流器向电池中注入恒定电流，直到电池充电到其容量的95%左右。”译为汉语的无主句，不过前提是不必或者不需要提及施动者，并在句子前添加“把”“使”等词，比如“Therefore, load is supplied from the AC line directly without any power conditioning. 因此，直接从交流线路为负载供电，不需要任何电源调节。”极少的情况下也可以

译为汉语的被动句,但并不是简单的译为"……被……"这种形式,而是借助汉语的表达被动意义的词汇,用主动的形式表达出来。这些处理方式都是比较符合汉语的使用习惯的。

翻译理论及应用讲评:图式理论

"图式"一词作为一个概念由德国古典哲学家康德(Kant)于18世纪哲学领域中使用,意指连接感知和概念的重要手段。图式理论则是由英国心理学家巴雷特(Frederick Barlett)于1932年提出和发展起来,他认为图式是对以往感知和经验的一种主动组织。后来皮亚杰(Piaget)将图式理论与认知论结合,他认为图式是认知的基础,具有动态性和可变性。而现代图式理论的代表人物鲁梅尔哈特(Rumelhart)1980年提出图式是以层级形式储存在长时记忆中的一组相互作用的知识和经验的结构或组块,其可构成认知能力。根据鲁梅尔哈特的图式理论,图式一般可分为:语言图式(linguistic schema)、内容图式(content schema)、形式图式(formal schema)。语言图式是对原语语言形式结构和语义结构的掌握程度,如词汇、语法和搭配等。内容图式是指关于事物、思维或现象的概念、背景知识等,包括了有关学科和专业方法的知识,以及其他语言知识、语用知识和百科知识。形式图式也称为结构图式,是对文本体裁和语篇结构的掌握程度,是有关文本的结构方法的背景知识,包括了有关语言的词法、句法和语篇等知识。

在本章原语文本中,语言图式的应用主要表现为大量的术语和专业名词短语的翻译。作为信息型文本,专业术语多是科技翻译的典型特征。科技术语的翻译应该符合内容精确、表达规范、简洁和理据性好。因此科技术语的翻译要忠于源语和译语的图式,用恰当的译语语言形式转译相关表达,积极努力实现与原文形式和功能的对等。比如电力领域有专门含义的"power distortion"(功率畸变),"inverter"(逆变器),"converter"(变换器),"rectifier"(整流器)等。另外还有一些是科技新词,起初是用来表示科技新事物、新现象、新规律,通过构词法(派生法、合成法和转换法)形成,比如"transformer"(变压器),"ferroresonance"(铁磁谐振),"low pass filter"(低通滤波器)。译者需要平时多学习掌握专业领域语言信息,积累丰富包括术语能力在内的专门用途科技英语能力。

内容图式在本章文本的翻译中发挥着重要作用。比如科技缩略词的大量使用反映了科技英语表达的简洁化趋势，这与汉字表意为主的特征有很大差异。比如“UPS”和“EMI”，完整词形分别是“uninterruptible power supply”和“electromagnetic interference”，译为“不间断电源”和“电磁干扰”。另外科技英语中还有大量使用名词化结构的特征，而且有许多为抽象名词和表静态倾向的名词。而科技汉语多使用动词，呈现动态倾向。所以有此内容图式，在翻译处理此类结构时就会适当运用词类转译的方法，进行词类转换，尤其是对表动作或状态的抽象名词或具有名词功能的动名词。如例⑤原句“Obviously, there is not a single configuration that can provide all of these features. Different configurations of UPS systems emphasize on some of the features mentioned above.”译文为“显然，一个单一配置的不间断电源系统无法具有上述所有特性，侧重要对上述某些特性差异化配置。”此处的两个“configuration”在译文中分别译为了名词词性和动词词性的“配置”。其中第一处“配置”译文之后增译了“系统”或“不间断电源系统”，而第二处则直接转译为动词，并省译了原文中的“UPS systems”。这样的处理避免了与上句内容的重复，体现了科技英语原文的客观、简洁的文体特征，又符合汉语的表达习惯，译文准确、通畅、简洁。

形式图式在“自下而上”与“自上而下”相互配合的原文理解过程有指导意义。“自下而上”的理解从单个字词句的理解到较大的语言单位的理解过渡，“自上而下”的理解过程则是译者利用各种图式，尤其是形式图式，从句群和篇章的整体含义，再到字词句的理解。“自下而上”和“自上而下”是交替进行的。“自下而上”的过程让译者可以了解句子结构和原文的内在组织形式，而“自上而下”的过程让译者理解句子之间的逻辑关系和语篇整体意义。比如科技英语是形合语言，空间结构呈树形或者葡萄式的扩展。科技英语长句中有大量的从句和非谓语形式和介词短语。而汉语重意合，呈线形护着竹节式的扩展，不用或少用逻辑连接词，句子间的逻辑关系更多依赖隐形连接。所以，科技英译汉对长句的“自上而下”的整体意义的理解很重要。具体针对长句的翻译，一般可先在理解句子整体意义的基础上，再通过“自下而上”地分析原句的主干，分析从属成分与主干成分的关系，为句内的词汇和短语寻找适当的语言表达形式。例①翻译难点分析表明了“从上而下”把握“power”一词在整个语篇和上下文语境中的词义选择的重要性。

另外，由于原语和译语的形式图式的差异，所以在正确理解原文意义的基础上，还应在译文中不要受限于原文的形式，而采取符合译语形式图式的表达。比如科技英语中的常用句型之一的被动语态，常用结构的分词短语结构和介词短语结构，连词“as”引导的句子，形式主语的句子都是典型结构和句式，而这些在汉语中并不能都可找到对应表达。因此翻译的时候，进行句型转换是必要的处理方法。如例④翻译难点分析表明了正确应用形式图式的差异对翻译英语被动语态的指导意义，因为在汉语中常用的是无主句和主动句，而不是英语中常用的被动句。

补充阅读：Transmission and Distribution Systems

In road networks, a range of designs are used depending on the circumstances. Freeways allow many cars to travel at once, with many lanes (so that travel is still possible even if a lane is closed due to an accident or maintenance), and with no interruption from traffic controls. Smaller regional roads that feed into and out of the freeways have traffic lights and lower carrying capacities. In the suburbs, lanes are even narrower and stop signs are common.

Power systems have similar characteristics. A useful distinction is between transmission and distribution subsystems.

1. Why We Need Different Voltages

Low voltages are used in general, low power applications as they hold a sufficient amount of energy for most consumer needs, without being too destructive during a fault.

Medium voltage and relatively high current is used in generation and loads to enhance the magnetic coupling. This allows the conversion of large amounts of power between mechanical and electrical form.

High voltage is used in distribution to allow bulk power to be distributed moderate distances with relatively little voltage drop. Distribution voltages allow economic plant configurations to be deployed over a wide area where a large number of equipment items need to be utilized. A relatively higher level of losses is tolerated to save on capital cost.

Extra high voltage is used in transmission to allow a high level of power transfer per ampere of current flowing through the line. As line losses are proportional to the square of the current and are not influenced by an increase in the system voltage, power can be transferred more efficiently, with fewer losses over the length of the transmission line per MW transferred. This means that at a higher system voltage, more power can be transferred for the same conductor size. This makes extra high voltage systems more economical for transferring large amounts of power. Transmission lines operating at extra high voltage form the backbone of a large power system. They are able to transfer large quantities of power over large distances, without losing excessive amounts of power in losses.

2. Transmission Circuits

A transmission circuit is a connection between two points (sometimes three) to allow the transfer of electrical energy. One circuit for a three-phase AC power system will have three conductors.

There is an advantage to have two points connected together by more than one electrical circuit, as this avoids loss of supply (or blackout) if one circuit has a fault or is out of service for maintenance.

A transmission circuit may be overhead or underground or a combination of both.

Sub-transmission systems distribute the power from load centres (or "points of supply") to the specific local area substations (also called "zone substations") where the power is further distributed.

Large power consumers, such as industrial sites or large commercial sites, may require their own zone substation, in which case they have a dedicated sub-transmission connection to their premises.

Distribution systems distribute the power from zone substations to roadside substations or other local reticulation, forming the final link to bring the power down to a voltage level where it is safe to use. Mostly, these systems are radial from each zone substation, with an open point between feeders of adjacent networks to allow transfer of loads between adjacent points of supply.

3. Lines

3.1 Overhead Line

An overhead transmission line uses steel towers, or wooden, concrete or metal poles, that hold the conductors at a safe distance above ground to avoid a short-circuit (or fault). Between the cross arm installed on the pole or tower and the electrical conductor, there are a stack of insulators. These prevent current from flowing to the pole and on to the ground, causing faults.

3.2 Cable

A cable surrounds each electrical conductor with an insulating material and a grounded conductive screen. This allows the cable to be installed in buildings or directly in the ground without the conductor short-circuiting.

3.3 HVDC

HYDC stands for High Voltage Direct Current. There are "classical HVDC" and "HVDC Light". The classical HVDC is useful for the transmission of large amounts of power, from a few hundreds of MW to about 3000 MW, over long distances. Complex equipment is required for the installation of an HVDC circuit, and does not compete economically with an AC transmission line over short distances. For transmitting high power over long distances (several hundred km) HVDC offers the following advantages:

- Only two conductors are needed (or even one conductor if the ground is used as return)
- Better system stability, i. e. two networks that are not synchronized or do not have the same frequency can be connected.

Chapter Seven Concept and Classification of Boiler

第七章 锅炉的概念和分类

原文

Boiler is a mechanical equipment which uses combustion heat or other heat of fuel to heat water (or other ③working medium) to produce steam, hot water (or other working medium, or other medium steam) with specified parameters and quality. Boiler used for power generation is called utility boiler or power plant boiler, or generally called steam generator.

译文

锅炉是指利用燃料燃烧的热能或其他热能来加热水(或其他工作介质)以生产规定参数和品质的蒸汽、热水(或其他工作介质、工质蒸汽)的机械设备。用于发电的锅炉被称为电站锅炉或电厂锅炉,又泛称为蒸汽发生器。

1 Basic Concept and Requirements of boiler

In the utility boiler, combustion heat comes from the combustion of fossil fuels (coal, petroleum, natural gas etc.) transfers to water (working medium) through metal wall of the heating surface, turning water into steam with certain pressure and temperature. Steam drives turbine and transforms heat energy to mechanical energy; then turbine drives power generator, and mechanical energy is transformed into electric power for users.

1 锅炉的基本概念和技术条件

在电站锅炉中,燃烧热来自化石燃料(煤、石油、天然气等)的燃烧,并通过加热面的金属壁向其中的水(工作介质或简称工质)传递,将水转化为具有一定压力和温度的蒸汽。蒸汽驱动汽轮机,将热能转化为机械能;随后,汽轮机带动发电机,机械能被转化为供给用户的电能。

The boiler in utility boiler means every heating surface that ③medium flows through, generally including ① economizer, water-cooled wall, superheater, reheater etc. and flow separation devices such as header, steam drum (steam water separator) etc. The furnace in utility boiler generally includes combustion place of fuels and smoke channel such as furnace, horizontal flue/gas duct and tail flue/gas duct etc.

电站锅炉中的“锅”，是指工质流经的每一个加热面，一般包括省煤器、水冷壁、过热器、再热器，以及联箱、汽包(汽水分离器)等分流装置。电站锅炉中的“炉”通常包括燃料燃烧场所及烟气通道，如炉膛、水平烟道及尾部烟道等。

Boiler parameters are the main indexes of boiler performance, including capacity of boiler, steam pressure, steam temperature, feed-water temperature etc. ④Capacity of boiler can be expressed by BECR or BMCR. BECR is the continuous production of steam per unit time with provision outlet pressure, temperature and efficiency. BMCR is the maximum continuous production of steam per unit time with provision outlet pressure, temperature.

锅炉参数是锅炉性能的主要指标，包括锅炉容量、蒸汽压力、蒸汽温度、给水温度等。锅炉容量可通过锅炉额定蒸发量(BECR, boiler rated evaporating capacity)或锅炉最大连续蒸发量(BMCR, boiler maximum continuous rating)来表示。锅炉额定蒸发量是指在规定的出口压力、温度、效率下，单位时间内连续生产的蒸汽量。锅炉最大连续蒸发量是指在规定的出口压力、温度下，单位时间内最大连续生产的蒸汽量。

Steam parameters include the pressure and temperature of boiler, and generally mean the pressure and temperature of superheated steam in superheater, reheater exit. If there are no superheater and reheater, steam parameters are the pressure and temperature of saturated steam in

蒸汽参数包括锅炉的蒸汽压力和温度，且通常指过热器、再热器出口的过热蒸汽压力和温度。如果没有过热器和再热器，蒸汽参数则是指在锅炉出口的饱和蒸汽的压力和温度。给水温度是指省煤器的进水温度，

boiler exit. Feed-water temperature is inlet water temperature of economizer. If there is no economizer, it refers to temperature of water entry into drum.

如果没有省煤器，则指汽包进水温度。

The overall structure of boiler includes two parts, boiler proper and auxiliary equipment. ⑤ Boiler proper is the hard core to produce steam which consists of furnace, drum, burner, water-cooled wall, superheater, economizer, air preheater, framework, brickwork and other main parts in boiler.

锅炉的整体结构包括锅炉本体和辅助设备两部分。锅炉本体是产生蒸汽的核心部分，它由炉膛、汽包、燃烧器、水冷壁、过热器、省煤器、空气预热器、构架、炉墙等其他主要锅炉部件组成。

The main requirements of utility boiler are safety, stabilization and economic. So the basic requirements of utility boiler are as follows.

电站锅炉的主要要求是安全、稳定、经济。因此，电站锅炉的基本要求如下：

① Evaporation capacity of boiler needs to satisfy the requirements of steam turbine generator set, running long-time in condition of nameplate parameters and having strong peak load regulation capacity. (2) Keeping normal temperature and pressure when running in wide load range. (3) Boiler should have high economical efficiency. (4) Using less steel, reducing initial investment and cost. (5) In the operation of boiler, it should have strong self-stable ability.

(1)锅炉蒸发量需要满足汽轮发电机组的要求，能在铭牌参数条件下长时间运行，并具有较强的调峰能力。(2)在宽负荷范围内运行时能保持正常温度和压力。(3)锅炉应具有较高的经济性。(4)少用钢材以减少初始投资，降低成本。(5)锅炉运行过程中，应具有较强的自稳能力。

2 Classification of Boiler

2 锅炉的分类

Boilers can be classified by circulation mode, combustion mode, slagging mode,

锅炉可以按照循环方式、燃烧方式、结渣方式、运行方式、燃

operation mode, fuel, steam parameter, furnace type and ventilation mode. The most common classifications are by circulation mode and steam parameter.

料、蒸汽参数、炉型和通风方式进行分类。最常用的分类方式是循环方式和蒸汽参数。

Circulation Mode

循环方式

In this way, boilers can be divided into natural circulation boiler, controlled circulation boiler and once-through boiler.

按这种分类方式,锅炉可以被分成自然循环锅炉、控制循环锅炉和直流锅炉。

A. Natural circulation boiler: Feed water comes into economizer through feed pump, and it is heated and then enters evaporation system. Evaporation system contains drum, unheated downcomer, heated water-cooled wall and corresponding header etc. ⑤ When feed water is heated in water-cooled wall, part of it will turn into steam. So medium in water-cooled wall is steam-water-mixture and in unheated downcomer is all water. Because water's density is larger than that of steam-water-mixture, pressure difference comes into being between downcomer and water-cooled wall. And pressure difference drives feed-water and steam-water-mixture to flow circularly in evaporation system. This circular flow is produced by heating water-cooled wall without any other energy consumption, so it is called natural circulation.

A. 自然循环锅炉:给水通过给水泵进入省煤器,加热后进入蒸发系统。蒸发系统由汽包、不受热下降管、受热水冷壁管和联箱等部分组成。当已加热的给水进入水冷壁管时,部分给水转变为蒸汽。因此,水冷壁管的工质是汽水混合物,而不受热的下降管部分的工质都是液态水。由于水的密度大于汽水混合物的密度,所以下降管和水冷壁管之间会产生压力差。并且压力差驱使给水和汽水混合物循环流动进入蒸发系统。这种循环流动由受热中水冷壁管产生,没有其他的能源消耗,因而被称为自然循环。

B. Controlled circulation boiler: By adding circulation pump into circulation

B. 控制循环锅炉:通过在循环回路加装循环泵,就可以增加

loop, flow driving force of medium can be increased to make controlled circulation boiler. ⑥In controlled circulation boiler, circular flow's pressure is stronger than natural circulation's, which brings more freedom to place water-cooled wall evaporation surface (which can be placed vertically or horizontally) and steam-water-mixture in it can flow up or down. This makes it meet the requirements of boiler structure better.

工质的流动推动力,形成控制循环锅炉。控制循环锅炉循环流动压力要比自然循环锅炉循环流动压力增强很多,因此可以更自由地布置水冷壁蒸发面,蒸发面可以竖直或者水平布置,并且汽水混合物可以向上或向下流动。这使得控制循环锅炉可以更好地适应锅炉结构的要求。

C. Once-through boiler: Once-through boiler has no steam drum and medium through evaporative part in one time and that means its circulation ratio is 1. Another characteristic of once-through boiler is there is no certain separation among economizer, evaporative part and superheater. Water all turns into steam in heating surface of evaporation and all flow resistance of flowing along medium is overcomed by feed pump. If adding circulating pump into once-through boiler start-up circuit, combined circulation boiler can be formed. That means when running below low load or ②Benson load, medium that passes surface of evaporation can't all turns into steam, and then saturated water separates from steam separator. The separated water reaches the entrance of economizer through circulating pump, and now medium that goes through evaporation part

C.直流锅炉:直流锅炉没有汽包,工质蒸发区是一次的,也就是说其循环倍率为1。直流锅炉的另一个特性是在省煤器、蒸发区和过热器之间没有固定的分界点。水都在蒸发受热面转变为蒸汽,沿着工质流动产生的所有流动阻力由给水泵解决。如果将循环泵加入直流锅炉启动回路,就形成了复合循环锅炉。这就意味着运行在低负荷或本生负荷状态时,由于工质流经蒸发器表面不能全部转变为蒸汽,饱和水会从汽水分离器中分离出来。被分离出来的水通过循环泵输送至省煤器的入口,这样进入蒸发区的工质流量多于蒸汽释放量,这使得循环倍率大于1。当锅炉负荷高于本生点或者以高负荷状态运行时,蒸发区产生微过热蒸汽,循环泵停

is more than steam releases, which makes circulation ratio > 1. When boiler runs above ② <u>Benson point</u> or at high load, evaporation part produces slightly superheated steam, circulating pump stops working and boiler works as pure once-through mode.	止工作,这样锅炉单纯以直流锅炉状态工作。
Once-through boiler is suitable for any pressure. But if pressure is too low, natural circulation boiler is better. So generally it's used in boiler of $P \geqslant 16$ MPa, and of course (ultra) supercritical parameters boiler can only use once-through boiler.	直流锅炉适用于任何工作压力。但是当工作压力过低时,自然循环锅炉性能更好。所以它通常用于 $P \geqslant 16$ 兆帕的锅炉,当然(超)超临界参数锅炉只可使用直流锅炉。

《术语表》

MPa	兆帕 megapascals
economizer	省煤器
water-cooled wall	水冷壁
superheater	过热器
reheater	再热器
furnace	炉膛
gas duct	烟道
drum	汽包
burner	燃烧器
downcomer	下降管
natural circulation boiler	自然循环锅炉
controlled circulation boiler	控制循环锅炉
once-through boiler	直流锅炉
circulating pump	循环泵

翻译难点讲评

① 对词语"Economizer"绝对不可望文生义,要是翻译成"节约者,节约装置,

经济家”就是错误的。查柯林斯词典得到其英文释义“a device that uses the waste heat from a boiler flue to preheat the feed water”。原来“economizer”是为了更有效地利用来自锅炉烟道的废气热量，将蒸汽锅炉所排放的废气导引通过这个设备，并在这个设备周围的管道系统继续流动循环，运用其余热来预热蒸汽锅炉。在百度百科查到一个词条，“省煤器——本词条由‘科普中国’科学百科词条编写与应用工作项目审核。省煤器（英文名称 Economizer）是安装于锅炉尾部烟道下部用于回收所排烟的余热的一种装置，是将锅炉给水加热成汽包压力下的饱和水的受热面，由于它吸收高温烟气的热量，降低了烟气的排烟温度，节省了能源，提高了效率，所以称之为省煤器。”

② 如果“Benson”按照一般专有名词人名的翻译方法，翻译成“班森（男子名）”的话就是错误的，因为专业领域的人名都是有实际含义的，有约定俗成的公认译法，也或者是固定译法。实际上 Benson 的全名是 Mark Benson，在网上搜该工程师的姓名，查到有一条关于“本生式直流锅炉和本生负荷”的介绍：“捷克工程师穆勒（一战期间到了英国，改名马克·本生 Mark Benson）在 1922 年发表了一项关于直流锅炉的专利，有别于汽包锅炉，这就是最早用于直流炉的型式。可以不再需要作为水汽分离装置的汽包，制造方便，这种结构利用了超过临界压力时水直接变成干蒸气的事实，具有布置在炉外的不受热下降管，尤其适用于大容量锅炉，这样便可免去蒸发时的两相状态，但是直流炉在刚开始启动时和汽包炉是一样的，只有达到了本生负荷的时候，锅炉的循环倍率为 1，这个时候的流量叫本生流量，其负荷也叫本生负荷。”

③ 对于“working medium”的简称“medium”是翻译成“工质”还是“介质”呢？这里涉及复合词术语的翻译和简写的问题。工作介质是各种热机中借以完成能量转化的媒介物质。这些物质通常以可逆的相变（如气-液相变）来增大功率，如蒸汽引擎中的蒸汽、制冷机中的雪种等。而介质是某种物质，如气体、液体、固体，这类能够被我们感受到的物质。电磁波、场等人直接感受不到的物质不是介质。介质（气体、液体、固体）中能溶解、通过、传播其他物质。对于后者而言，前者称为后者的介质。如对盐、糖而言水就是溶解它的介质，对声音，空气、固体就是传播它的介质。介质也叫做媒介（物）。由此可以得出“working medium”的简称“medium”要翻译成“工质”。

④ 通过搜索网络，查到“BECR”的意思可以是“额定蒸发量”“蒸汽流量”“额

定工况”和“额定负荷”，而“BMCR”的意思可以是“锅炉最大连续蒸发量”和“锅炉最大连续出力”。考虑到BECR的完整形式是“boiler rated evaporating capacity”，BMCR的完整形式是“boiler maximum continuous rating”。以及原文中的上下文“BECR is the continuous production of steam per unit time with provision outlet pressure, temperature and efficiency. BMCR is the maximum continuous production of steam per unit time with provision outlet pressure, temperature.”所以，分别翻译成“额定蒸发量”和“锅炉最大连续蒸发量”较好。另外，需要注意的是科技缩略语可能有多个意思，比如BECR就还可以指“Bureau of External Cultural Relations（中）文化部对外文化联络局”。因此对于缩略语的翻译一定要选择专业领域相关的，与上下文有关的词义。

⑤ 本处原文有多处人称照应。它是用人称代词或相应的限定词所表示的照应关系。由于人称代词及相应的限定词在英语中作为篇章纽带的使用率，远远大于在汉语中的使用率，所以要避免机械地照搬，应注意三点：a.英语强调代词照应时，如指代清楚无歧义，则译文中应省略不译，可着重于语气之连贯与文辞之简约及文句之紧凑。b.为了连贯起见，可将在译文中产生歧义的代词还原成它所替代的名词。c.一个从句在主句前的英语复合句在使用名词和代词时，其次序在大多数情况下是和汉语一致的，即：初提用名词，复指用代词，但英语中人称代词有时出现在被替代的对象之前，即“名词放在主句，代词放在从句”；而汉语中人称代词总是出现在所指对象之后，照应上文的那个对象。汉译时，可把原文中的代词及其所指对象的位置对换，使所指对象出现在代词之前，代词就有了明确的归属。比如，“part of it will turn into steam”这里的代词“it”所指的应该是“feed water 给水”，翻译的时候还原为所指对象。“in unheated downcomer is all water”之前应该省略了共同成分“medium”，所以翻译的时候进行了还原，增译出“工质”。“that of steam-water-mixture”这里的“that”指的是“density”，对此进行明确，还原为“密度”。当然也不是所有的指代成分都需要增补出来，比如“so it is called natural circulation.”这里的“it”就可以省译，省译后更加简洁，也不会影响语义。这个句子“Boiler proper is the hard core to produce steam which consists of...”里面的“which”所指的“boiler proper”，是第二次提到，就可以在译文中放在后面用代词“它”替代。

翻译理论及应用讲评:翻译伦理

伴随着20世纪90年代翻译研究的范式从规定转向描写,译者伦理的研究也开始回归。这一时期,韦努蒂(Venuti)1995年的 *The Translators' Invisibility—— A History of Translation*,皮姆(Pym)1997年的 *Pour une Ethique du Traducteur*,切斯特曼(Andrew Chesterman)1997年的 *Means of Transaltion: the Spread of Ideas in Translation History* 等著作对翻译伦理都有涉及。2001年《译者》(The Translator)特刊第二期名为"The Return to Ethics"提出从伦理角度研究翻译基本问题的重要意义。在该特刊中切斯特曼的文章"Proposal for a Hieronymic Oath"提出了翻译的四种伦理模式,即再现伦理(ethics of representation)、服务伦理(ethics of service)、交际伦理(ethics of communication)和规范伦理(norm-based ethics)。再现伦理即对原文文本内容和风格的忠实再现,译本准确再现原作文本和文化基本特征,译文和原文之间最大限度地对等。服务伦理是基于商业和服务性质的翻译活动而提出的,即译者接受委托人的提议和要求,译者对客户负责,同时还要考虑不违反原作者的意图和译文读者的期待。这个观点与功能主义目的论的主张基本一致。交际伦理实质上是要求译者不被动地再现原文作者的意图,而是主动承担起作者、读者之间交流沟通的任务,尽力让译语读者了解原文的意义。基于规范的伦理,即满足特定文化的社会道德和伦理规范,在译语文化中符合国家与社会的期待。

下面举例说明本章文本的翻译中如何体现再现伦理和交际伦理。侧重分析例⑥中对连贯的再现。原文"In controlled circulation boiler, circular flow's pressure is stronger than natural circulation's, which brings more freedom to place water-cooled wall evaporation surface (which can be placed vertically or horizontally) and steam-water-mixture in it can flow up or down."译文为"控制循环锅炉循环流动压力要比自然循环锅炉循环流动压力增强很多,因此可以更自由地布置水冷壁蒸发面,蒸发面可以竖直或者水平布置,并且汽水混合物可以向上或向下流动。"本例中原文使用两个"which"引导词,标明了其定语从句的性质,一个用逗号与先行词隔开,另一个用括号与先行词隔开。并且最后一个小句中"...and...in it..."用并列连词和代词与句子前面的内容照应。在英语中,上面提到的几个都是衔接手段(照应/替代/省略/逻辑连接/词汇衔接),对句子的连贯性不可或缺,否则就违反了基本语言规则。而汉语代词用得少,汉语往往不用连接词表达句间或者句内逻辑关系,因为汉语倾向于使用词汇重复的手段表达衔接关系。所以,将该例中的英语译为汉语,其中的衔接手段可加以转变,以符合汉语的衔接规则。汉语的译文把"circulation's"后面省略的内容都补足完整了,把本来是介词短语作状语的"In controlled circulation boiler"省略介词后与原主句主语合并,这样两句中就出现了词汇重复的衔接。两个"which"的原始功能在译文中都被弃用。第一个"which"转换为了表因果关系的"因此",第二个"which"重译为"蒸发面",分别与前文形成逻辑衔接和词汇重复的衔接。最后"and"表并列关系在译文中该功能保留不变,但是省去了此后的"in it",因为此时译文句子的主语已经是"控制循环锅炉"。此外"in it 在其中"的意思是不言自明的。可见译文再现了原文文本的连贯,又传递了全部的意义,可谓"准确、通畅"两者兼得。

可见,再现伦理代表的是对原文的忠实,译文要体现原文表达的全部内容,这是对应"真实"的价值观的。再现"真实"即对原文作者的最大忠实。而且必须看到,由于语言、文化的不同,译本不会是原文的翻版。因为译本要经过译者对原文的理解,用符合译语语言文化和规范的形式表达出来。译者为了实践交际的伦理,在充当沟通的桥梁时,必须尽力消除信息流通的障碍。在这个过程中,译者可根据具体的文本,采取适合的翻译策略和翻译方法。

补充阅读:Substation Equipments: Transformer, Circuit Breakers, Bus and Disconnector

A substation is a point on the power system that connects multiple circuits and may also change the voltage to more useable levels. The common uses for a substation are:

• connection of a generator to a transmission network by means of a step-up transformer;

• connection of a customer network by means of a step-down transformer;

• the ability to share power between multiple electrical circuits;

• in the event of a fault on a connected circuit, faults can be cleared and power redistributed therefore minimizing disruption to power supply

• planned disconnection of circuits to allow maintenance to occur on power system equipment, without disrupting power supply.

A substation may consist solely of a transformer and two switches mounted on a pole. A main substation on a particular network, however, is more likely to be a site complete with control room, indoor and/or outdoor switchgear and transformers.

1. Transformer

Transformers allow the voltage level of a network to be increased to transfer large amounts of power over long distances, and then bring the voltage level back down to a level useable by consumers.

Transformers serve two purposes: they can transfer power across two systems of different voltages; and/or they can provide electrical isolation between electrical systems.

A two-winding transformer can connect a network of one voltage level to the primary winding, and one network to the secondary winding. Transformers commonly contain all three phases in the same enclosure, so there are three primary windings (one for each phase), and three secondary windings.

The ratio between the number of turns on the primary and secondary windings determines the voltage conversion ratio.

As transmission voltages can vary depending on network loading, the voltage conversion ratio of a transformer can be fine-tuned using a tap changer. A tap changer can increase or decrease the number of turns on one side of a transformer to change the voltage conversion ratio. The significance of this is that the voltage of the primary network can vary while the transformer regulates the voltage of the secondary network. The electrical isolation between the primary and the secondary network is achieved by having no direct electrical connection between the conductors on the primary and secondary sides. Furthermore, the magnetic coupling resists changes in the amount of current flowing (presented as an impedance), which limits the magnitude of fault current. There are three main types of transformers that are used for power transfer between transmission networks.

• Generator step-up (or generator transformer). A generator transformer is made to accept the large currents generated at medium voltage, and transfer the power to the transmission system at a high voltage level and relatively low current. Generator transformers typically have a large voltage ratio and have a delta connection on the generator side to limit the flow of zero sequence current, which limits the generator earth fault level.

• Interconnection transformer. An interconnection transformer transfers power between transmission networks of different voltages. These transformers have a small voltage ratio.

• Step-down transformer. A step-down transformer reduces the voltage from one level to another. There are two main categories for step-down transformers:

• Supply (zone substation) transformer — this steps the voltage down from transmission or sub-transmission to distribution voltage levels.

• Distribution transformer — steps the voltage down from distribution high voltage to low voltage.

Other types of transformers used in the power systems include earthing transformers, current transformers and voltage transformers.

2. Circuit Breakers, Bus and Disconnector

Circuit breakers (usually abbreviated as "CB") are switches which are used commonly in substations that are capable of breaking fault current. In the event of a fault, circuit breakers can be switched very rapidly to minimize the amount of damage that occurs at the fault location and all connected equipment, as well as minimizing the disturbance to customers.

During fault conditions, it is likely that large currents (possibly due to a short-circuit) will flow through the circuit breaker. In order to successfully clear the fault, the circuit breaker must be designed to be able to open the switch (or break the connection) during this maximum current flow.

The bus describes the electrical conductor that connects all the circuits together in a substation. It allows the current from any connected circuit to flow to any other circuit. The current flow through the bus cannot be directly controlled; however, the switching of equipment can change the current flow.

Different bus configurations are used in different circumstances.

Disconnectors provide an essential, visible, electrical isolation gap between two points on an electrical network.

If maintenance was required on a transmission line, theoretically, it would be sufficient to open the circuit breakers at each end and link a connection to earth on each side of the maintenance crew for safety. However, circuit breakers, unlike disconnectors, do not show a visible break in the network, and do not adequately demonstrate to local personnel that one section of the network is isolated.

Disconnectors also provide a mechanism for isolating both terminals of the circuit breakers so that maintenance can be carried out on the circuit breaker itself. Disconnectors can also allow the quick reinstatement of all non-faulted equipment after a fault has occurred. This is particularly important when dealing with circuit breaker failure.

Chapter Eight General Safety Requirements of Electrical Maintenance or Repairs

第八章 电气设备维护与维修基本安全要求

原文

This section deals with the reliability and effective maintenance of electrical systems that can be achieved in part by careful planning and proper design. The training of personnel in safety-related work practices that pertain to their respective job assignments is outlined.

Only qualified persons shall perform electrical repairs. It is dangerous for an unqualified worker to attempt electrical repair. Before any electrical maintenance or troubleshooting is performed, sources of electrical energy shall be deenergized, except where it is necessary for troubleshooting, testing, or areas that are infeasible to deenergize. All energy sources shall be brought to a safe state. ⑧ For example, capacitors shall be discharged and high capacitance elements shall be short-circuited and grounded.

译文

本节介绍电气系统的可靠性和有效维护。可以通过精心规划和正确设计在一定程度上实现电气系统的可靠性和有效维护。本节也概述了与安全相关的实际工作中相关工作人员的培训。

只有具备资质的人员才能进行电气维修。无资质人员尝试进行电气维修工作是很危险的。除在故障排除或测试需要，或所在地区无法切断电源的情况下，在电气系统的维护或故障排除前应切断电源。所有电源都需处于安全状态下。例如，电容器都应放电，高电容元件都应短路并接地。

1 Work on Energized/Deenergized Electrical Equipment

⑨The first consideration for working on any electrical system is to have the circuit positively deenergized. ③All circuits and equipment must be considered energized until opened, tagged and/or locked according to an approved procedure ⑤and should be proven deenergized by testing with an approved testing device known to be in proper working order , and have to be checked by r eview ing system drawings and/or perform ing system walkdowns. ⑦Where the possibility exists that the circuit can become energized by another source or where capacitive devices (including cables) may retain or build up a charge, the circuit should be grounded and shorted. The grounding and shorting device should be selected and installed in accordance with appropriate standards. ⑩ Whenever work is to be performed on a positively deenergized system, the worker must also identify and protect against any accidental contact with any exposed energized parts in the vicinity of the work.

1 带电/断电电器设备作业

任何电气系统作业都应当首先考虑电路确认断电。根据已批准程序,除线路开路、设备被挂牌或锁闭的情况外,所有线路和设备都应被视为带电。应使用经批准的工作状态良好的检测设备来验证线路和设备已断电。应检查系统图纸并/或进行系统实地巡查。如果有其他电源电路仍然可能被通电,或如果电容设备(包括电缆)仍带电或能够产生电荷,电路就应接地并短路。应按照适当的标准选择和安装接地及短路设备。即使系统确认已断电,在开展工作时工作人员仍需要确认并防止意外接触作业区域周围暴露在外的任何带电部件。

2 Considerations for Working on Energized Systems and Equipment

Qualified employees performing such tasks as electrical repairs, modifications,

2 带电系统及设备作业的注意事项

有资质的员工在带电系统、部件或设备上进行电气维修、改

and tests on energized electrical systems, parts, and equipment need to comply with the following:

(1)⑦Live parts to which an employee may be exposed shall be deenergized before the employee works on or near them, unless the employer can demonstrate that deenergizing introduces additional or increased hazards or is infeasible due to equipment design or operational limitations.

(2)Work performed on energized electrical systems and equipment may be done only if a supervisor and/or cognizant safety professional and the personnel performing the work determine that it can be done safely. Approval should be given for each job. Approval for the same job performed repeatedly may be given through the use of an approved procedure or job safety analysis.

(3)⑪ Personnel shall not work on energized circuits unless they are qualified to do so, or, for training purposes, unless they work under the direct supervision of a qualified person.

(4)Sufficient protection in the form of insulated tools and insulated protective equipment, such as gloves, blankets, sleeves, mats, etc., shall be used while working on energized circuits.

Note: The above discussion assumes

装或测试时,应当遵守下列规范:

(1)除非雇主能够证明断电将增加额外风险,或由于设备设计及操作限制而无法断电,在工作人员进行作业或位于带电部件附近前,应将可能暴露的带电部件断电。

(2)只有管理人员和/或安全监督专业人员以及工作人员确认可以安全施工后,才可在带电系统或设备上开展工作。各项工作都需经过批准。通过使用已批准的程序或工作安全分析来批准重复开展的同一工作。

(3)除非人员本身具备资质或者进行培训时如果没有具备资质人员的直接监督,禁止在带电电路上开展工作。

(4)进行带电线路作业时,可以使用绝缘手套、毯子、袖套、垫子等绝缘工具和绝缘保护设备进行充分的防护。

注:上文中的讨论假设系统

the system voltage is less than the maximum use voltage of the ASTM class of rubber goods used.

(5) Other work, ① independent of voltage, ② that presents a significant shock or arc blast hazard to employees, needs to be evaluated as to the number of employees involved.

(6) At least two employees shall be present while the following types of work are being performed:

(A) Installation, removal, or repair of lines that are energized at more than 600 volts.

(B) Installation, removal, or repair of deenergized lines if ④ an employee is exposed to contact with other parts energized at more than 600 volts.

(C) Installation, removal, or repair of equipment, such as transformers, capacitors, and regulators, if ④ an employee is exposed to contact with parts energized at more than 600 volts.

(D) Work involving the use of mechanical equipment, other than insulated aerial lifts, near parts energized at more than 600 volts.

(E) Other work that exposes ④ an employee to electrical hazards greater than or equal to those listed above.

Exceptions to the items listed above

电压低于美国材料试验协会(ASTM, American Society For Testing And Materials)标准的橡胶制品的最大使用电压。

(5)其他工作,虽不受电压影响,但如对工作人员带来大冲击或电弧爆炸等危害,需要根据涉及的工作人员的人数做出评估。

(6)执行以下类型的操作时,至少应有两名工作人员在场:

(A)高于600V带电线路的安装、移除或维修。

(B)工作人员对断电线路进行安装、移除或维修时接触其他高于600V的带电零件。

(C)安装、移除或维修设备,如变压器、电容器及调节器时,工作人员接触其他高于600V的带电零件。

(D)除绝缘架空升降机外,在高于600V的带电零件附近使用机械设备作业时。

(E)任何其他将工作人员置于上述所列同等或比之更大的电气危险中的作业时。

上述所列项目的例外情况:

are:

(F) Routine switching of circuits, if the employer can demonstrate that conditions at the site allow this work to be performed safely.

(F)如果雇主单位能证明现场条件能保证施工安全时,进行常规的电路切换。

(G) Work performed with live-line tools if ④ the employee is positioned so that he or she is neither within reach of nor otherwise exposed to contact with energized parts, and

(G)工作人员所处位置不在带电零件区域内,并且不与带电零件接触时,使用带电线路工具作业。

(H) Emergency repairs to the extent necessary to safeguard the general public.

(H)为保障广大市民安全进行的必要紧急维修。

(7) Taking voltage measurements may subject personnel to exposed energized parts. Where it is determined personnel are subject to contacting exposed energized parts, personnel shall use the appropriate protective equipment for the voltage levels involved.

(7)测量电压时电工可能暴露在含带电零件的环境下。与暴露的带电零件接触时,工作人员需根据电压等级使用适当的保护设备。

3 Safety Watch Responsibilities and Qualifications

3 安全监管职责与资格

The responsibilities and qualifications of personnel for sites that require the use of a safety watch are as follows:

安全监管人员的职责及资质如下:

(1) Trained in cardiopulmonary resuscitation (CPR);

(1)接受过心肺复苏术(CPR)培训;

(2) Thorough knowledge of the locations of emergency-shutdown push buttons and power disconnects in their operations;

(2)全面了解操作中紧急关闭按钮及切断电源的位置;

(3) Thorough knowledge of the specific working procedures to be followed

(3)全面了解应遵循的具体工作程序及应完成的工作;

and the work to be done;

(4)⑥Specific responsibilities include monitoring the work area for unsafe conditions or work practices and taking necessary action to ensure abatement of the unsafe condition or work practice, deenergizing equipment and alerting emergency-rescue personnel as conditions warrant, maintaining visual and audible contact with personnel performing the work, and removal of inju red personnel, if possible;

(4)具体负责检测工作区域的不安全情况或不安全作业，采取必要行动确保消除不安全情况或不安全作业，情况允许时切断设备电源并警报紧急救援人员，保持与工作人员的音视频联系，如有可能，转移受伤人员；

(5) The safety watch should have no other duties that preclude observing and rendering aid if necessary.

(5)如有必要，安全监管人员可不担负其他职责，以免妨碍监管及提供援助。

《术语表》

troubleshooting	排除故障，故障检修
deenergize	去激励；断电
capacitor	电容器
ground	接地；地线
walkdown	巡查
short	短路
live part	带电部件；有电部件
energized circuit	带电线路
ASTM class	美国材料试验协会标准
live line	带电线路
safety watch	安全监管人员

翻译难点讲评

① 由于英汉两种语言的差异，有时英语句中的某个词或短语在翻译时无法转换成对应的词或短语，只好将其扩充为分句，同时对语序也作相应的调整。在需要扩展成句的英语词类中，占比例最高的是副词，其次是形容

词,还包括少量的动词。英语副词需要扩展的比例较高,大概是因为英语副词使用较多,与其他词类的搭配较灵活,而汉语则没有对应的表达方式。这个时候可以把词和短语升格为短句,比如:"independent of voltage",这个插入成分本也可以翻译成主语名词前的定语,但是考虑到它在句中的角色除了定语,还隐藏着其他的逻辑关系,这里可以和下面的定语从句一起处理,它可以升格为让步状语从句的一个短句"虽不受电压影响"。

② 这个整句较长,句子成分或者主从句之间逻辑联系密切。但是往往也不必囿于个别词句的意思而忽视了译文的隐藏逻辑关联。比如"Other work, independent of voltage, that presents..."这里把原来的定语从句和前面的形容词短语插入语可构成一个汉语的"虽然……但是"的让步状语从句,定语从句和后面的主句是暗含的"因为……所以"因果关系,所以最后翻译为"其他工作,虽不受电压影响,但如对工作人员带来大冲击或电弧爆炸等危害,需要根据涉及的工作人员的人数做出评估。"译文对原文表达方式进行调整后,更加符合汉语逻辑表达方式。

③ 增译是常见的翻译方法。增词的情况主要分为:原文省略了某些词汇而不影响句子意义,但翻译时为保证句子意思需要补全省略的词汇;原文某种意思寓于字里行间,译成汉语时需要通过增词表达隐含意义。这个句子这个部分"until opened, tagged and/or locked"因为省略了与前文重复的内容,英文表达非常简洁。但是为了明晰化的需求,在译文中需要增译省略成分,"除线路开路、设备被挂牌或锁闭的情况外"。

④ 由于中、英语言的差异,有些单词在原文中必不可少,但是若全部展现在译文中却会造成译文的冗余拗口。为保证译文的自然流畅,在翻译时可以根据汉语的语言规范省略这种词语。这里下划线的几个带不定冠词和定冠词的"employee"和"employer"都非指实际数量的概念,所以只需翻译为"工作人员"和"雇主单位"。

⑤ 原文句子结构译文无法复制,或复制不符合汉语习惯时所采用的一种译法是分译,或曰拆译,即将原文拆成若干段,使其中的一个短语或一个从句译成汉语的一个分句 。分译时,必须考虑被分拆的各个分句保持本身意思的相对完整,以及前后分句间的逻辑联系。必要时,添某些汉语词语,把隐含的意思充分表达出来。常见的有:a.原句的谓语前拆开译成分句;b.原句的介词前拆开译成分句;c.原句的分词前拆开译成分句;d.原

句的关系代词、关系副词前拆开译成分句；e.原句的连词、副词前拆开译成分句；f.原句的两端拆开，中间抽出译成分句；g.将原句的结构灵活拆开，重新组合译成分句。比如该句中有三个并列的谓语动词“must be considered...”“should be proven”和“have to be checked”。翻译的时候在谓语动词的地方分译成为三个句子。拆开后后面的两个句子可以不翻译为被动句，而翻译为汉语的无主句。

⑥ 无生命主语句指用无生命的事物或抽象名词做主语的句子，是英语书面用语的典型句式，汉语中无生命主语较少。这个下划线句子中的谓语动词均与人（安全监管人员）有关，翻译时重新调整句子结构，改为有生命主语句表达更顺畅。具体的方法就是把对原来的主语“Specific responsibilities”进行转换，名词转译为动词“负责”，相应的原来的形容词定语转译为副词“具体”，原来的谓语动词“include”只能省译。

⑦ 语篇的衔接手段分为词汇衔接以及指称、省略、替代、连接等四种语法衔接。英语中第三人称代词使用比汉语更为频繁，因此对于英语第三人称指称，翻成汉语时常改为省略、替代或原词复现等衔接方法。这里下划线句子“Live parts to which an employee may be exposed shall be deenergized before the employee works on or near them”中的“them”在译文中进行了还原，译为“带电部件”。连接词在文章中表达了作者组织信息的逻辑以及语义上的轻重缓急。“通过连接性词语的运用，人们可以了解句子之间的语义联系，甚至可经前句从逻辑上预见后续句的语义。”（胡壮麟，1994）。连接词既是一种语法手段，又是一种词汇手段。这里下划线句子英语原文“Where the possibility exists that the circuit can become energized by another source or where capacitive devices (including cables) may retain or build up a charge, the circuit should be grounded and shorted.”中用“where”状语从句，实则表示条件，译文则采用了连接词“如果……就”做衔接。如果此处译文同样采用地点状语的关系就会显得比较奇怪，而将地点状语关系变为条件关系，译文表达更顺畅，但是加重了原文的语气。

翻译理论及应用讲评：卡特福德翻译转换理论

翻译转换的研究受到了20世纪50—60年代结构主义语言学的影响。其中1965年卡特福德的翻译转换模式最为著名，这在其专著《翻译的语言学理论》(*A Linguistic Theory of Translation*)第十二章有专章讨论。除了卡特福德之外，其他学者，如波波维奇(Popvi，1970)，鲁文·兹瓦特(Leuven Zwart，1984，1989，1991)，布朗姆·库尔卡(Blum Kulka，1986)和图里(Toury，1995)都曾提出了各自的翻译转换模式，但适用性和可操作性皆不如卡氏的模式。比如，鲁文·兹瓦特的翻译转换模式有37种转换形式，是最详尽、最全面的模式，但是需要花大量的学习时间，很难对翻译实践有直接的指导作用。卡氏的模式包含两种转换类型，即层次转换(level shifts)和范畴转换(category shifts)。

卡特福德翻译转换是从原语到译语形式对应的偏离，结果可以做到文本等值/对等(textual equivalence)。层次转换发生在原语与译语不存在同一语言层次(linguistic levels)上的对应语言表达形式，但具有不同语言层次上的译语翻译等值成分。常见的是从语法与词汇层次之间的互相转换，比如英语动词的时态和语态和语气，可数名词的复数变化在汉语中都不存在对应的词语变化对应形式。此时可转换为汉语的词语，尽管形式不对应，但是功能上可明确表达原文时态、语态和语气。根据不同等级上或者等级之间的形式对应的脱离(departures from formal correspondence)，范畴转换可分为四种：结构转换(structure shift)，可发生在所有等级里，如主被动语态转换，肯定否定结构转换，英语主语突出与汉语的主题突出转换等；词类转换(class shifts)发生在英汉词语等级上不同类别的词之间的相互转换；单位转换(unit shifts)发生在英汉语不同等级之间的等值成分相互转换，如在句子、子句、短语、单词四种语言单位之间的转换；内部体系转换(intra-system shifts)反映了在同一体系内部的同一等级上与原语形式对应的译语形式转换为译语同一体系内的同一等级上的其他形式，比如与原语有大致对应形式的译语转换为译语内非对应的形式，这经常发生在如数字体系、指称体系、动词时体形式、动词语态、动词语气和科技新词的翻译中。

例⑧的翻译过程体现了层次转换和结构转换。原文“For example, capacitors shall be discharged and high capacitance elements shall be short-circuited and grounded.”译文为“例如，电容器都应放电，高电容元件都应短路并接地。”此句的两个复数主语名词在汉语中没有复数形式对应，但可以增加一个副词“都”，放在谓语动词之前，可表达出“全部，所有的”复数概念，并与上一句中的“所有电源都需处于安全状态下”相互照应。原句谓语动词的两处被动形式都译为汉语中的主动形式。原句中的动作发出者不需要出现，译文用主动在语义上贴近原文，表达上更符合汉语习惯。

例⑨的翻译过程体现了词类转换。原文“The first consideration for working on any electrical system is to have the circuit positively deenergized.”译文为“任何电气系统作业都应当首先考虑电路确认断电。”此句的两个具有动词含义但是分别为名词词性或名词功能的“consideration”和“working”都转换为译文中的动词词性。此类转换也可发生在副词、形容词、动词等词类之间。

例⑩的翻译过程体现了单位转换。原文“Whenever work is to be performed on a positively deenergized system, the worker must also identify and protect against any accidental contact with any exposed energized parts in the vicinity of the work.”译文为“即使系统确认已断电，在开展工作时工作人员仍需要确认并防止意外接触作业区域周围附近暴露在外的任何带电部件。”此句中的短语“a positively deenergized system”本来是时间状语从句复合谓语里动词后面的宾语成分。但在译文中该短语转换为一个让步状语从句“即使系统确定已断电，……”，并放在原来的时间状语从句的位置之前，以对此内容加以强调。在确切表达原义的同时，这样的转换既简化了原文的结构，又符合汉语多用小短句的趋向，更便于读者阅读和理解。

例⑪的翻译过程体现了内部体系转换。原文“Personnel shall not work on energized circuits unless they are qualified to do so, or, for training purposes, unless they work under the direct supervision of a qualified person.”译文为“除非人员本身具备资质或者进行培训时如果没有具备资质人员的直接监督，禁止在带电电路上开展工作。”此句中有两个“unless”，第一个和第二个“unless”在句子内限定的范围有所区别，而且这也导致两者在句内的语义也有不同，所以在译文中也有必要体现出来此不同。尽管“unless”本来的词义是“除非”，第二个却翻译为如果没有。译文中使用“除非”是与“unless”直接对应的译文，译文中使用“如果没有”或者“假如没有”体现了“除非”这个语义在汉语语言系统内部的系统转换，该转换体现了对原文内容的客观阐释。

补充阅读:Substation Equipment: Earth Switch, Current/ Voltage Transformer

1. Earth Switch

An earth switch allows the earthing of parts of a circuit, as a safety precaution during maintenance. The earth switch will protect against inadvertent re-livening of an electrical circuit during maintenance; and will provide a path for any stray currents that may flow if the circuit is in close proximity to other live circuits.

2. Current/Voltage Transformer

Instrument transformers form the eyes and ears of the power system. These devices allow the safe measurement of current and voltage present on the main conductors. As indicated by the name, these measurement devices transform the electrical property they are measuring down to a level suitable for input into electronic devices as well as insulating the measuring devices from the power system. These devices feed information to relays which monitor the system; and to the network operator, so they can manage the power flow.

A current transformer generally uses the primary conductor as the primary winding. It uses a secondary coil that is wired around a toroidal core that is positioned around the main conductor to measure the current. The number of turns of the secondary coil determines the current reduction ratio; the ratio is chosen to reduce normal operating current down to a level that protection equipment can use to make measurements.

An open circuit of the secondary winding of a current transformer is very dangerous as the secondary winding acts as a current source. This means that the secondary winding would try to 'drive' the current through whatever material forms the open point of the circuit. It could deliver a fatal level of current to personnel in close proximity, as well as damaging the insulation on the secondary winding.

A voltage transformer operates similarly to a step-down power transformer, in that it reduces the voltage from any system voltage (i. e. 6.6 kV—330 kV) down to 110 V, However, the main difference with a voltage transformer is that it is not designed for power transfer, and is only designed to transfer the minimum amount of power necessary to make an accurate voltage measurement (typically 100—200 YA).

Chapter Nine Permanent Magnet Motor

第九章 永磁电动机

原文

A synchronous motor is a type of AC motor in which the shaft rotation is in synchronization with the frequency of the supply current. The stator of the motor creates a magnetic field which rotates in time in exact synchronization with the line frequency. Permanent magnet motor is a type of synchronous motor in which permanent magnets are embedded in the rotors which creates fixed magnetic fields.

译文

同步电动机是一种交流电动机，其中轴的旋转频率与供电电流的频率同步。电动机的定子产生旋转磁场，其转速与电网频率是及时精确同步的。永磁电动机是一种同步电动机，其中永磁体嵌入在转子中，产生固定磁场。

1 Overview

The ① permanent magnet synchronous motor (PMSM) can be thought of a bridge between ① brushless DC (BLDC) motor and AC induction motors. In PMSM the rotor structure containing the permanent magnet is very similar to the brushless DC motors and the stator structure resembles closely with that of AC induction motor.

A three phase permanent magnet motor having one permanent magnet pole rotor is shown in figure 1.1. A constant

1 概述

永磁同步电动机(PMSM)可以认为是无刷直流电动机和交流感应电动机之间的桥梁，在永磁同步电动机中，包含永磁体的转子的结构与无刷直流电动机十分类似，而定子的结构则与交流感应电动机相类似。

图 1.1 为具有一个永磁极转子的三相永磁电动机。(图片省略，后同)由嵌入电动机旋转

magnitude rotor flux is created by the permanent magnet which is embedded in the rotating axis of the motor, the stator winding which is controlled by the stator current creates rotating electromagnetic field when energized. The torque is produced by the intersection of rotor and stator fluxes and this torque developed makes the motor to rotate and it is represented in the figure 1.2. ⑥By efficiently controlling the angle between the stator and rotor magnetic fluxes, maximum torque and maximum electromechanical conversion efficiency can be achieved.

轴的永磁体产生恒定的转子磁通量，由定子电流控制的定子绕组在通电时产生旋转电磁场。由转子和定子磁通的交点产生转矩，并且这个生成的转矩使电动机转动，如图1.2所示。通过有效地控制定子和转子磁通量之间的角度，可以产生最大转矩和实现最大机电转换效率。

Depending upon the rated speed and power requirements of the machine there are various rotor structure topologies, also the rotor mechanical structure depends on the number of poles and desired flux gradients. ③Up to few kilowatts of power required, permanent magnet is used in synchronous machines to generate the magnetic field, for higher power rotor windings are used and in these winding DC current is circulated so that magnetic field is generated in the windings. The rotating stator field must rotate at the same frequency as the rotor permanent magnetic field, otherwise a rapidly changing positive and negative torque is experienced by the rotating rotor, which leads to mechanical vibration resulting in wear and tears of

由于电动机的额定速度和功率要求，存在多种转子结构拓扑，转子机械结构也取决于极数和期望的通量梯度。需求高达几千瓦的功率时，同步电动机中使用永磁体来产生磁场，由于使用的更高功率的转子绕组，并且在这些绕组中循环直流电流，从而绕组中产生磁场。旋转的定子场必须以与转子永磁场相同的频率旋转，否则旋转的转子会经历快速变化的正负转矩，这导致机械振动使得机械部件磨损和破裂，进而引起该转矩小于可以产生的最佳转矩。在这些条件下，电动机也可能经历失步现象，此时转子在同步频率停止旋转并且静止转子参考坐标系中

the mechanical parts and also less than optimum torque is produced. The machine under these conditions may also experience the ② pull out phenomena where rotor stops rotating at the synchronous frequency and average torque in the stationary rotor reference ②frame is zero.

的平均转矩为零。

2 Permanent Magnet Motor Drive System

2 永磁电动机驱动系统

In a motor drive system, power converter and controller are very important aspect and ⑦ all of motor drive system disciplines revolve around machine design, power converter design and controller design and implementation. The general block diagram of motor drive system is shown in figure 1.3.

As shown in the block diagram inverter and controller unit are very important for the motor drive system. Position sensor is used to detect the rotor position. Rotor position is very important as for maximum torque generation it is desired to keep the stator and rotor fluxes orthogonal to each other.

在电动机驱动系统中，功率转换器和功率控制器是非常重要的方面。所有电动机驱动系统规范都围绕机器设计，功率转换器设计和功率控制器设计以及它们的实现来进行。电动机驱动系统的总体框图如图 1.3 所示。

如框图所示，逆变器和控制器元件对电动机驱动系统非常重要。位置传感器用于检测转子位置，转子位置对于产生最大转矩非常重要，因为该位置能够保持定子绕组电流方向和转子磁场方向相互正交。

3 Inverter

3 逆变器

A voltage source inverter is generally used to feed the motor with the current. The voltage source inverter converts the ①DC source voltage to AC voltage. For

电压源逆变器通常用于为电动机提供电流。电压源逆变器将直流电压转换为交流电压。为了实现电动机运行电压频率

achieving variable frequency and magnitude of the voltage for motor operation generally first ① AC voltage is converted to DC and that voltage is fed to inverter to get the AC voltage for the motor input. The inverter output current which goes to the stator of the motor is regulated and the gating signal for the switching devices in the arm of the inverter is generated.

和幅值可调，通常首先将AC电压变换为DC电压，并且该DC电压馈送到逆变器以获得用于电动机输入的AC电压。调节流向电动机定子逆变器的输出电流，并产生逆变器换流桥中开关器件的选通信号。

The voltage source inverter shown in figure 1.4 has 6 switches and anti-parallel diode is also shown. ⑧ The inductive loads require path for the flow of the current when the switch is turned off and that path is provided by the anti-parallel diodes also termed as free-wheeling diode. The switching devices and the diodes are selected depending upon the power and current rating of the inverter. Most commonly used devices for switch are ① bipolar junction transistors (BJT), metal oxide semiconductor field effect transistors (MOSFET) and insulated gate bipolar transistors (IGBT). MOSFET is widely used in the low power and low voltage applications, IGBT is more commonly used in motor drives and industrial applications for medium and high power range.

图1.4所示的电压源逆变器有反向并联二极管和6个开关。开关断开时，反向并联二极管提供感应负荷需要电流流过的通路，因此该反向并联二极管也称为续流二极管。开关器件和二极管的选择取决于逆变器的额定功率和额定电流。最常用的开关器件是双极结型晶体管(BJT)，金属氧化物半导体场效应晶体管（简称场效应晶体管，MOSFET）和绝缘栅双极型晶体管(IGBT)。场效应晶体管在低功率和低电压应用中得到广泛应用。绝缘栅双极型晶体管在中等和高功率的电动机驱动器和工业应用中更为常用。

The switching sequence and the gating signals for the voltage source inverter are shown in figure 1.5. In figure 1.6 the line to neutral voltage and line to

电压源逆变器的开关顺序和选通信号如图1.5所示。图1.6显示了六步电压源逆变器的线电压。由于线路与中点电

line voltage for a six step voltage source inverter is shown. The inverter is termed as six step voltage source inverter due to the presence of six steps in the waveform of the line to neutral voltage.

压的波形存在六个步骤,因此该逆变器被称为六步电压源逆变器。

The line to line voltage for the six step voltage source inverter is given by the following equations. ④From the equations we can observe that for the voltage source inverter, the output amplitude can be controlled by the input DC voltage (V_{dc}).

六步电压源逆变器的线电压由以下公式给出。由公式可以看出,电压源逆变器的输出幅度可以由输入直流电压(V_{dc})控制。

The three most basic control schemes being implemented in the voltage source inverter are as follows: hysteresis band current control, sinusoidal pulse width modulation (SPWM), space vector pulse-width modulation (SVPWM).

在电压源逆变器中实现的三个最基本的控制方案如下:滞环电流控制,正弦脉宽调制(SPWM),空间矢量脉宽调制(SVPWM)。

In hysteresis control an external reference is given and the main purpose of the control is to make the ac line current follow the given reference. Hysteresis control loop for current in phase A is shown in figure. Whenever the actual current i_a goes beyond the given reference, then the status of the switches in the phase a leg (S_1 & S_4) changes.

在滞环控制中给出了外部参考,其控制的主要目的是使交流线路电流跟随给定的参考值。图中显示了A相电流的滞环控制回路。每当实际电流 i_a 超出给定参考值时,A相桥臂(S_1 和 S_4)中开关的状态就会改变。

In the figure, hysteresis current control waveform is shown, the change in the switching state can be observed when the current i_a goes beyond the reference. This controller can be simply implemented by using an operational amplifier.

在图中,显示了迟滞电流控制的波形,当电流 i_a 超出参考值时,可以观察到开关状态的变化。该控制器可以通过使用运算放大器来实现。

The hysteresis current control has very good dynamic response and it's easy to implement at very low cost. On the other hand the hysteresis control has large current ripple in steady state and modulation process generates the sub harmonic components. ⑤ In hysteresis controller, there is no inter relation between the three controllers used for the three phases of current and hence there is no strategy to generate zero voltage vectors and hence the switching frequency increases at lower modulation index and signals leaves the hysteresis band as the zero vector is turned on.

滞环电流控制具有非常好的动态响应,并且很容易以非常低的成本实施。另一方面,滞环控制在稳定状态下具有大电流纹波,并且在调制过程中会产生次谐波分量。因为在滞环控制器中,用于三相电流控制的三个控制器之间没有相互关系,所以没有产生零电压矢量的对应策略。因此,在调制指数较低时开关频率会增加,信号通道在零矢量过程打开保留下迟滞环带。

《术语表》

stator	定子
rotor	转子
synchronous motor	同步电动机
permanent magnet motor	永磁电动机
brushless DC motor	无刷直流电动机
AC induction motor	交流感应电动机
winding	绕组
torque	转矩,扭矩
rotor flux	转子磁通
electrom conversion	机电转化
flux gradient	通量梯度
position sensor	位置传感器
voltage source inverter	电压源逆变器
gating signal	选通信号
free-wheeling diode	续流二极管

anti-parallel diode	反向并联二极管
MOSFET	金属氧化物半导体场效应晶体管
IGBT	绝缘栅双极型晶体管
hysteresis control	滞环控制
pulse width modulation	脉冲宽度调制

翻译难点讲评

① 缩略词是将复杂的英文单词简单化的结果,包括已经被行业或社会认可的缩略词,也包括作者为了使行文流畅自然而自创的缩略词。在对缩略词进行翻译时,应确定其缩略前的全称,再附上缩略语。若同样的缩略语反复出现,翻译一次即可。PMSM 放在括号里面,是为了方便下文出现 PMSM 时直接使用。所以,在进行翻译时要保留原文格式。

再看 DC 和 AC,二者分别表示直流(direct current)和交流(alternating current),出现在"brushless DC motor"和"AC induction motors"中。因为二者并不独立,若保留缩略词,汉语则译为"无刷直流(DC)电动机"和"交流(AC)感应电动机"。这样的译文不具有可读性,连贯性差。所以在翻译时将其对应全称的汉语意思译出即可,DC 和 AC 可不作保留。

下划线句子里的"MOSFET"全称为"metal oxide semiconductor field effect transistors"。查询百度百科有一词条"场效应晶体管(field effect transistor,缩写 FET),简称场效应管。主要有两种类型(junction FET, JFET)和金属氧化物半导体场效应晶体管(metal-oxide semiconductor FET,简称 MOSFET)。"但是在上下文语境中会直接将"MOSEFT"简称为"场效应晶体管",所以可以采取同样方式进行处理。第二句中再次出现"MOSFET"时,将其处理为"场效应晶体管",既尊重原作者,也对读者负责。

② 有些词虽是普通词汇,但在专业领域使用的时候却又不同于普通词义。对同一词汇在专业领域的不同译法要区别不同的情况,一是在上下文语境相同的情况下对词汇的恰当理解及翻译,二是在不同的语言环境下词汇的具体含义,这就需要联系上下文语境来确定词义,三是要根据前后搭配,四是要结合专业领域作专业化的引申。比如此处下划线的"frame",在不同的专业领域里面有以下词义:"边框［建］巢框［蜂］车架［车辆］刚架［力］构架［机］(伞、自行车等的)骨架［轻］画幅调节［电影］画格

［电影］画面［电影］机架［机］机座［机］架［科技］架子［科技］结构［科技］壳框［古生］框架［建、计］肋骨［船］模套［塑料］排字台［印刷］全架［公路］筛框［冶］托架［机］悬架［机］（植物）御寒玻璃罩［农］帧［计、电子、通信］支架［建］”。如果没有“rotor”，直接查“stationary reference frame”，可以找到其译文是“静止参考坐标系”。其实如果对于“reference frame”这个搭配比较熟悉的，可以很快判断出其原来是和“reference”一起搭配表示“参考坐标系”的。可见每个词在一个新的语境中就是一个新的词。

同理，这里的“pull out”绝非是“离开，撤离；拔出；渡过难关；恢复健康”的意思。它与 phenomena 一起搭配，表示专业词义“使不同步，使失步”。在百度知道网站上有这样的留言可以验证：“发电机失步指的是发电机或电力系统振荡开始时过剩力矩很大，转子惯量使发电机的工作点不断向功角增大方向移动，一直冲过功率极限点，汽轮机的输入功率与发电机功率无法平衡。实际运行中，造成失步的主要原因有：系统发生短路故障；发电机励磁系统故障引起发电机失磁，励磁不足（欠励磁）使发电机电动势剧降；发电机电动势过低或功率因素过高；系统电压过于低于额定值；负载力矩过大；电机接收到的信号失真。”可见此处应该翻译为“失步现象”，保证了译文的忠实和专业准确性。

③ 科技英语陈述客观规律和原理、反映客观现象和变化、描述客观过程和事实，突出“客观”两字，力戒主观意念和个人好恶（方梦之，2005:96），所以在科技英语中的句子中，最常见的就是被动句。在本句中，共出现了五个被动形式，如果翻译成“被需求”“被使用”“被用于”“被循环”“被产生”，根据目的论的连贯原则，应注意译文表达的得体，直接保留汉语中的“被”，读起来会拗口，所以翻译时选择将英语的被动句变为汉语的主动句或者无主句。

④ 这里的句子“From the equations we can observe that for the voltage source inverter, the output amplitude can be controlled by the input DC voltage (Vdc).”主句主语是“we”。若是碰到这种情况，不能盖棺定论认为该文本不够客观。近一二十年来科技英语也有多用主动态的倾向，认为主动态表达清晰、明确、亲切（方梦之，2005:96）。虽然使用第一人称做主语不太符合科技英语的规范，但可以根据目的论的目的性原则，翻译的目的决定了翻译的方法，为了使译入语文本依然保持科技英语的规范，译者可以在

条件允许的情况下，将主语改译或省去。这里的译文“由公式可以看出，电压源逆变器的输出幅度可以由输入直流电压(Vdc)控制。”中，省去了原来的“we”，直接使用了无主句。这样做虽然改变了译入语文本的特征，但是符合汉语表达习惯。

⑤ 科技翻译强调逻辑性和条理性，而句子间的衔接和连贯正是逻辑性和条理性的保证。衔接是指一段话中个部分在语法或词汇方面有联系或两方面都有联系。这种联系可能存在于不同的句子之间，也可能存在于一个句子的几个部分之间(方梦之，2005：116)。根据目的论，译文无须与原文完全对等，为了明晰化的交际目的，不应拘泥于原文格式，应具体情况具体分析，根据不同的翻译目的，选择不同的翻译方法。可以顺译，也可对原文进行拆分后分译，以及删减或者添加相应的词。该长句的三个组成句子之间用的是“and hence”连接在一起。仔细分析，可知前两个小句之间是一层因果关系“In hysteresis controller, there is no inter relation between the three controllers used for the three phases of current and hence there is no strategy to generate zero voltage vectors”，前两个小句合起来和第三个小句构成因果关系“and hence the switching frequency increases at lower modulation index and signals leaves the hysteresis band as the zero vector is turned on.”。所以前面增译了关联词“因为在滞环控制器中，用于三相电流控制的三个控制器之间没有相互关系，所以没有产生零电压矢量的对应策略。”后面分译成一个单独的句子“因此，在调制指数较低时开关频率会增加，信号通道在零矢量过程打开保留下迟滞环带。”

翻译理论及应用讲评：翻译补偿理论

翻译补偿就是为了减少翻译损失，而翻译过程中信息、意义、语用功能、文化因素、审美形式的损失有不可避免性。因此，以目的语规约为主，辅之以其他语言手段，根据翻译目的和文本类型，对翻译损失进行修复或者弥补是尤其重要的。

翻译补偿作为理论命题的研究开始于20世纪80年代，在此之前只是作为一种翻译实践中的翻译方法和翻译理论家译论中的翻译补偿思想存在。国内的学者80年代以后的20年间，关注的是文化差异所致的翻译损失的补偿策略和具体的补偿手段,如柯平(1991),屠国元(1996),孙迎春(1996)和刘树森(1997)诸多学者提出的增词法、引申法、拆译法、融合法、替代法、转移法、视点转换、具体化、概略化等手段。21世纪初国内学者对于翻译补偿的研究不再局限于具体方法的研究,呈现跨学科的趋势,如从语言学和哲学角度研究补偿问题,也有马红军等在借鉴国外学者研究的基础上进行翻译补偿类型的研究。2006年夏廷德的专著《翻译补偿研究》代表了国内学者首次把翻译补偿进行理论体系的建构。之后的研究继续了在补偿策略和文化补偿上的扎堆进行研究,同时学者们也出现了从阐释学、模因论、顺应论、关联理论、图式理论、功能翻译理论等多学科多视角地建构补偿理论体系。

国外对于补偿和相关论述最早始于20世纪60—70年代,如Nida and Taber(1969),以及Wilss(1982),但是相关论述不成体系,也不明确。Newmark(1988)对补偿的功能和分类的限定相对更为清晰,但在翻译损失覆盖范围上太窄,后来最著名的论述来自Hervey and Higgins(1992:35-40)的*Thinking Translation: A Course in Translation Method*的四大分类法:转类补偿、转位补偿、兼并补偿和拆分补偿。

夏廷德(2006)就是以此四大分类为基础提出了八大分类法:同类补偿、异类补偿、原位补偿、异位补偿、整合补偿、分立补偿、同步补偿和差异补偿。夏廷德为此提出的翻译补偿原则有六种:需求原则、相关原则、重点原则、就近原则、等功能原则、一致原则,即分别指有原语损失为补偿前提,补偿内容与原文和原作者的意旨相关,依据文本类型和翻译目的重点补偿翻译损失,语篇层面实施异位补偿以实现整体效果等值,按照原文形式负载的功能进行补偿,以及保证译文和原文内容和整体风格的一致。

翻译补偿在具体操作上一般有两个层面,即语言学层面上的词汇补偿、语法补偿和语篇补偿,以及审美层面上的审美形式的功能、价值冲突和形义统一性等方面的补偿。科技翻译过程中的翻译损失也同样需要翻译补偿,以下从本章文本和译文中举例加以说明。

科技翻译由于要达到表达通顺流畅的标准,以及词汇层面英汉两种语言系统固有的差异,务必对转换中发生的词汇的语义损失采用增译、具体化、概略化、加注、阐释、归化、视点转换和回译的方法准确流畅地表达出来。如例⑥原文“By efficiently controlling the angle between the stator and rotor magnetic fluxes, maximum torque and maximum electromechanical conversion efficiency can be achieved.”译文为“通过有效地控制定子和转子磁通量之间的角度，可以产生最大转矩和实现最大机电转换效率。”此句中的充当谓语动词的“achieved”与句子的两个名词短语和主语成分之间可看作是多枝共干的被动关系。谓语动词“achieved”的词义笼统，担当的是弱势动词的功能。所以该词在译文中进行了具体化的处理，分别与两个名词短语重新组合为主动的动宾搭配，分别具体化为“产生”和“实现”两个词义。

例⑦原文“... all of motor drive system disciplines revolve around machine design, power converter design and controller design and implementation.”译文为“所有电动机驱动系统规范都围绕机器设计，功率转换器设计和功率控制器设计以及它们的实现来进行。”此句译文增译了原文中的许多省略成分。比如原句中“controller design”前省略了“power”，“implementation”前省略了“their”。这在科技英语中反映了其简洁、客观的特征，但在译文中这些省略成分不翻译出来的时候，容易影响到语义表达。比如如果译为“……和控制器设计以及实现”，这样的译文准确对应了原句的形式，但明显会发生语义损失，因此此处还是补偿出省略的成分才能表达更准确。

另外例⑦中也有语法补偿。语法补偿是因为科技英语和汉语的语法方面的差异很大，主要体现在动词的时/体/态/式，名词的单复数等。比如此例中的“disciplines”是复数可数名词，译文中补偿了一个汉字“都”，该字原句没有，但是可表达出名词的复数含义。

语篇补偿主要指的是逻辑关系和指称关系。比如例⑧原文“The inductive loads require path for the flow of the current when the switch is turned off and that path is provided by the anti-parallel diodes also termed as free-wheeling diode.”译文为“开关断开时，反向并联二极管提供感应负荷需要电流流过的通路，因此该反向并联二极管也称为续流二极管。”该句的语义重心在于说明开关断开时反向并联二极管的独特功能，这也是为何它被称为续流二极管的原因。译文补偿了该层隐藏的逻辑因果关系。另外，原句的两处语法被动形式都转换为汉语主动句形式。还有原句中“that path”这处代词+名词的短语的指代与其指代内容“...path for...”之间有很长的介词短语和时间状语割裂开。汉语译文中把这层有指代关系的两处内容进行了融合处理，使得它们距离拉近，表义更清楚易懂。

补充阅读:Substation Equipment: Surge Arrestors, Reactive Power Support, Earthing

1. Surge Arrestors

Surge arrestors are a form of 'pressure release, for a power system. They allow current to flow to earth through them if the voltage becomes too high. Surge arrestors are useful in all kinds of damage. A surge arrestor is not the same as a short-circuit; when a surge arrestor allows current to flow to earth it does not reduce the voltage level of the connected electrical network below its nominal operating value.

2. Reactive Power Support

Reactive power gives the power system the voltage support it needs to transfer real power from generation to loads.

Reactive power support can be installed at any point on a network (although substations are commonly chosen) to help stabilise the surrounding network and ensure normal operation. Common types of reactive power support include:

- Capacitor Bank. At times of large loading on a power system at points that are distant to generation, the voltage can drop to undesirable operational levels. In extreme cases it can lead to unplanned load tripping.

The net effect of installing capacitors is a boost in voltage, which helps to alleviate voltage sag during heavy loading. Voltage sage is caused by an increase In the reactive power flow in the supply circuit, and a subsequent increase in the voltage drop along the length of the circuit. Capacitors can also cause voltage rise if the reactive power flow is reversed.

- Shunt Reactor. A shunt reactor can be required to prevent over-voltages from occurring at parts of the network, either during periods of light loading or when a supply point in the network rises with the export of large volumes of power.

- Static Var Compensator (SVC). An SVC allows a variable amount of reactive power to be supplied or consumed which is consequently used to control the voltage of a point on the network to a desired level over a range of different operating scenarios.

3. Earthing

At most substations all equipment with metallic parts, which are not part of the main electrical circuit, are earthed to ensure safety and provide a common reference for the power system

At large substations the physical connection to the ground is usually made by buried copper conductors connected together in grid like fashion. In smaller installations or for transmission poles the earthing connection can consist of a single pin driven into the ground.

Chapter Ten Turn-On Check-Out of DC Power Supply

第十章 直流电源开机检查

原文

1 Before Check-Out

② Ensure that the power supply is configured as follows:

The unit is connected to an appropriate AC source.

The POWER switch is in the off position.

⑥ Sense connector pins 1 and 2 are jumped; sense connector pins 4 and 5 are jumped.

⑥ All switches on Connector J2 are in the down position.

SHOCK HAZARD Be aware that hazardous voltages can be present on the output terminals. ② Do not set the output voltage above 40 VDC during the turn-on check-out procedure.

2 Constant Voltage Check

(1) Turn the POWER switch on.

(2) ③ Turn the output on by pressing the OUT ON button. ⑤ The green OUT ON indicator should be illuminated.

译文

1 在检查之前

必须确保电源配置如下：

本机连接到合适的交流电源。

POWER 开关处于关闭位置。

传感器连接器接头引脚 1 和 2 断开；传感器连接器接头引脚 4 和 5 断开。

连接器 J2 上的所有开关处于断路位置。

电击危险，请注意，输出端子上可能存在危险电压。在开机检查过程中，请务必不要将输出电压设置在直流 40 伏以上。

2 恒压检查

（1）打开电源（POWER）开关。

（2）按下开启输出（OUT ON）按钮打开输出，绿色的 OUT

ON 指示灯应发亮。

(3) ⑤ The green CV indicator should also be illuminated. ② If the CC indicator is illuminated, rotate the current knob until the CV indicator becomes illuminated.

(3)绿色恒压(CV)指示灯也应该亮起。如果恒流(CC)指示灯亮起,请旋转电流旋钮,直到恒压(CV)指示灯亮起。

(4) ⑦ Rotate the voltage knob while observing the DC VOLTS display.

(4)旋转电压旋钮,观察直流电压(DC VOLTS)输出显示屏。

(5) The output voltage should vary while the knob is turned. The voltage range is from zero to the maximum rated output for the power supply model.

(5)旋钮旋转时,输出电压应该变化。电压范围从零到电源输出模式的最大额定输出。

3 OVP Check

3 过压保护检查

(1) Rotate the voltage knob and set the output voltage of the unit to 50% of its full-scale rating or 30 volts, whichever is lower.

(1)旋转电压旋钮,将本机的输出电压设置为最大额定值的50%或30伏,以较低者为准。

(2) ⑦ Press the OVP/UVL button once so that the DC AMPS display indicates OVP. ① The DC VOLTS display shows the OVP level.

(2)按下过压保护/欠压锁定(OVP/UVL)按钮一次,直流电流(DC AMPS)显示屏应显示 OVP (overvoltage protection)。直流电压(DC VOLTS)显示屏应显示 OVP 电压值。

(3) Use the voltage knob and set the OVP level of the unit to 75% of its full-scale voltage rating or 40 volts, whichever is lower.

(3)旋转电压旋钮,将本机的 OVP 电压设置为最大额定电压的75%或40伏,取较低者。

(4) ① Wait a few seconds until the DC VOLTS display returns to show the output voltage.

(4)等待几秒钟,直到 DC VOLTS 显示屏重新显示输出电压值。

(5) Use the voltage knob and raise the output voltage of the unit until it

(5)旋转电压旋钮提高设备的输出电压,直到接近过压保护

approaches the OVP setting. Check to make sure that the output voltage cannot be set higher than the OVP setting.

设定值。检查输出电压以确保输出电压不高于过压保护设定值。

(6) Press the OVP/UVL button again. Rotate the voltage knob and reset the OVP level of the unit to its maximum setting.

(6)再次按下 OVP/UVL 按钮。旋转电压旋钮并将本机的过压保护电压重置为其最大设定值。

4 UVL Check

4 欠压锁定检查

(1) ⑧ Press the OVP/UVL button twice so that the DC AMPS display indicates UVL. The DC VOLTS display shows the UVL level.

(1)按下 OVP/UVL 按钮两次，DC AMPS 显示屏应显示 UVL(undervoltage lockdown)电压值。直流电压显示屏应显示 UVL 电压值。

(2) ③ Use the voltage knob and set the UVL level of the unit to 50% of its full-scale voltage rating or 30 volts, whichever is lower.

(2)旋转电压旋钮并将设备的 UVL 电压设置为最大额定值电压的50%或30伏，取较低者。

(3) Wait a few seconds until the DC VOLTS display returns to show the output voltage.

(3)等待几秒钟，直到 DC VOLTS 显示屏重新显示输出电压值。

(4) Use the voltage knob and lower the output voltage of the unit until it approaches the UVL setting. Check to make sure that the output voltage cannot be set lower than the UVL setting.

(4)旋转电压旋钮并降低设备的输出电压，直至达到欠压锁定电压设定值。检查输出电压以确保输出电压不能低于欠压锁定电压设定值。

(5) Press the OVP/UVL button twice. Rotate the voltage knob and reset the UVL level of the unit to its minimum setting.

(5)按下 OVP/UVL 按钮两次。旋转电压旋钮并将设备欠压锁定电压重置为其最小设定值。

5 Constant Current Check

(1) Turn the POWER switch off. Wait a few seconds until the AC indicator on the front panel goes out.

(2) Use a heavy wire and short the +V and -V output terminals together.

(3) Turn the POWER switch on.

(4) ③ Turn the output on by pressing the OUT ON button. The green OUT ON indicator should be illuminated. The green CC indicator should be also illuminated.

(5) Rotate the current knob while observing the DC AMPS display. The output current should vary while the knob is turned. The current range is from zero to the maximum rated output for the power supply model.

6 OCP Check

(1) Rotate the current knob and set the current limit of the unit to about 10% of its full-scale current rating.

(2) Press the OCP/488 button. ④ This should trip the OCP protection.

(3) The OCP indicator should be illuminated, the DCVOLTS display should indicate OCP, and the alarm indicator should be blinking.

(4) Press the OCP/488 button again to cancel OCP protection. The DC VOLTS

5 恒流检查

(1) 关闭电源开关。等待几秒钟直至前控制板的交流信号指示器关闭。

(2) 使用粗导线将电压正极和负极输出端短接起来。

(3) 打开电源开关。

(4) 按下“OUT ON”按钮打开输出。此时，绿色的开启输出指示器会亮起，绿色的恒流指示器也会亮起来。

(5) 在观察 DC AMPS 显示器的同时旋转电流旋钮。随着旋钮的旋转输出电流会随之变化。电流变化范围是从零到电源输出型号的最大额定输出。

6 过流保护检查

(1)旋转电流旋钮，将电源设备的电流极限设置为其最大额定电流的约10%。

(2) 按下过流保护(OCP/448)按钮。该动作会触发过流保护。

(3) OCP 指示器会亮起来，DC VOLTS显示器会显示 OCP，与此同时警报指示器会闪烁。

(4) 再次按下 OCP/488 按钮以关闭过流保护。由于过流

display should indicate OFF because the OCP protection is latched.

(5) Press the OUT ON button to reset the OCP protection. The output should return to its previous setting.

(6) Turn the POWER switch off.

(7) Remove the short from the + V and − V output terminals.

7 Normal Operation

The power supply has two basic operating modes: constant voltage and constant current mode. In constant voltage mode, the power supply regulates the output voltage at the selected value, ③ while the load current varies as required by the load. In constant current mode, the power supply regulates the output current at the selected value, ④ while the voltage varies as required by the load. The mode in which the power supply operates at any given time depends on the voltage setting, current limit setting, and the load resistance.

8 Constant Voltage Mode

When the power supply is operating in constant voltage mode, the CV indicator on the front panel illuminates. Adjustment of the output voltage can be made when the output is enabled (ON) or disabled (OFF). When the output is enabled, sim-

保护已经锁住，DC VOLTS 显示器会显示 OFF 状态。

(5)按下 OUT ON 按钮来重置过电流保护。此时，输出会回到之前的设置。

(6)关闭电源开关。

(7)移除电压正极和负极输出端的短接线。

7 一般工作模式

电源有两种基本工作模式：恒压模式和恒流模式。在恒压模式中，电源调节输出电压在选定值，负载电流随负载的不同需求而变化。在恒流模式中，电源调节输出电流在选定值，此时电压随负载的不同需求而变化。在任何给定时间内，电源运行的模式取决于电压设置、电流极限设置和负载电阻。

8 恒压模式

当电源运行在恒压模式时，位于前控制板的 CV 指示器亮起。在启用或禁用输出电压时，均可以调整输出电压。输出启用时，只需旋转电压旋钮来设定输出电压。在输出关闭时，先按

ply rotate the voltage knob to program the output voltage. When the output is disabled, press the LIMIT button and then rotate the voltage knob. The DC VOLTS display will show the programmed voltage for 5 seconds after the adjustment has been completed and then indicate OFF. The voltage knob can be set to coarse or fine resolution. Press the FINE button to select finer resolution. The FINE indicator turns on.

下极限(LIMIT)按钮然后再旋转电压旋钮。当调节完成后，DC VOLTS 显示器将会显示5秒钟设定的电压，之后会显示OFF状态。电压旋钮可以设置为低或高的解析率。按下微调(FINE)按钮来选择更高的解析率，此时FINE指示器会开启。

①If you cannot adjust the voltage to the value that you desire, the power supply may be operating at its current limit. Check the load condition and the current limit setting. Also, the voltage cannot be programmed lower than about 5% above the UVL setting, or higher than about 5% below the OVP setting.

如果你未能将电压调到想要的值，那么电源有可能运行在电流极限状态。检查一下负载的情况和电流极限设置。另外，电压不能设定得低于欠压保护设置的约5%，也不能高于过压保护设置的约5%。

9 Constant Current Mode

9 恒流模式

When the power supply is operating in constant current mode, the CC indicator on the front panel illuminates.

电源工作在恒流模式，前端面板上的CC指示灯亮起。

④Adjustment of the output current limit can be made when the output is enabled (ON) or disabled (OFF). When the output is enabled and in constant current mode, simply rotates the current knob to program the current limit. If the output is in constant voltage mode, press the LIMIT button and then rotate the current knob. The DC AMPS display will

当输出开启(ON)或关闭(OFF)时，可以调整输出电流极限。当输出开启并处于恒流模式时，只需要旋转电流旋钮即可调节电流极限。如果输出处在恒压模式时，则需要先按下LIMIT按钮，再旋转电流旋钮。在输出电流极限的调整完成后，DC AMPS显示屏将显示5秒钟

show the programmed current for 5 seconds after the adjustment has been completed and then indicate the actual output current.

的设定电流值，然后显示实际输出电流。

When the output is disabled, press the LIMIT button and then rotate the current knob. ① The DC AMPS display will show the programmed current for 5 seconds after the adjustment has been completed and then goes blank because the output is off.

当输出禁用，按下 LIMIT 按钮并旋转电流旋钮。在输出电流极限的调整完成后，DC AMPS 显示屏将显示 5 秒钟的设定电流值，随后由于输出被关闭，设定电流值上将变为空白。

The current knob can be set to coarse or fine resolution. ①Press the FINE button to select finer resolution. The FINE indicator turns on.

电流旋钮可以设置为粗解析率和高解析率。按下 FINE 按钮以选择较高解析率，并且 FINE 显示器开启。

术语表

VDC	直流电压
OUT ON	开启输出
CV	恒压
CC	恒流
DC VOLTS	直流电压
OVP	过压保护
UVL	欠压极限
DC AMPS	直流电流
OCP	过流保护
FINE	微调
LIMIT	极限
resolution	解析

翻译难点讲评

① 增译法就是在原文的基础上添加必要的单词、词组、分句或完整句，从而

使译文在语法、语言形式上符合汉语的习惯并使译文在修辞上、语法结构上、词义上或语气上与原文保持一致，达到译文与原文在内容、形式和精神方面对等的目的。这里的增加不是无中生有地随意增加，而是增加原文中虽无其词却有其意的词语。虽然从理论上说可以根据具体的上下文增加词，但在什么时候增加什么词，才能恰到好处，而不超出一定的界限，则需要悉心体会。运用增译法对科技英语译文进行处理，应遵循一定的原则，即：统一性原则（如名词复数“-s，-es”的翻译）；具体性原则（如抽象名词后增译名词“情况、作用、现象、性、效应、过程、装置、设计、变化”使其具体化）；完整性原则（增译量词、省略成分、关联词、概括性词语和表时态的时间副词或助词）；准确性原则（如隐含意义的明晰化）；合理原则（如被动语态时增译表达生命主体的名词或代词和祈使句时增译“命令、请求”的词）；解释性原则（如对专有名词和文化负载词的解释性增译）。

如此句“Wait a few seconds until the DC VOLTS display returns to show the output voltage.”的“display”和“voltage”需要根据准确性原则，不能只翻译成“显示”和“电压”，而应该译为“显示屏”和“电压值”。同理，这句“The DC AMPS display will show the programmed current for 5 seconds after the adjustment has been completed and then goes blank because the output is off.”中的“adjustment”也需要根据准确性原则把隐含的意思明晰化或根据完整性原则把省略成分补充出来，把它译为“输出电流极限的调整”，意思更为明确，因为该句所在段落前面的句子中有“adjustment of the output current limit”。句子“If you cannot adjust the voltage to the value that you desire, the power supply may be operating at its current limit.”中的“if”根据完整性的原则，需要翻译为汉语里的“如果……那么”。此外，该句中的“current limit”和下划线其余句子里的“fine resolution”和“OVP level”，不能简单译为“电流极限，高解析和过压保护电压”，而应该根据具体性原则，增补相应的名词，分别译为“电流极限状态，高解析率和过压保护电压值”。

② 英语的祈使句具有“命令、请求”或“叮嘱”等含义，汉译时常常根据对象、场合，增补一些词语如“请、要、应、须、千万、一定、务必”等，以符合汉语表达习惯。如以下句子和它们的翻译：Ensure that the power supply is configured as follows（必须确保电源配置如下）；Do not set the output voltage above 40 VDC during the turn-on check-out procedure.（在开机检查过程

中，请务必不要将输出电压设置在直流40伏以上。)；If the CC indicator is illuminated, rotate the current knob until the CV indicator becomes illuminated.[如果恒流(CC)指示灯亮起，请旋转电流旋钮，直到恒压(CV)指示灯亮起。]

③ 由于英汉两种语言的表达方式和语言结构的差异，在翻译时，就需要根据意义和句法的需要在原文的基础上增添一些必要的词或减少一些多余的词，使译文通顺、流畅、自然，并且更加符合汉语的表达习惯。因此，在忠实于原文的基础上，适当地采用增译或省译这两种翻译方法可以提高科技英语的翻译质量。在科技文献中，经常遇到某些词语的省译。

如果逐词直译，会造成译文不通顺，使读者无法理解。但是省译不是任意删改英语原文，要在完整表达英语原文内容的前提下，把译文写得言简意赅，读起来像是典型的汉语写作。

a.冠词的省译：英语中使用冠词很多，但有不少是没有什么意义的，可以省译。但在一些常用英语词组中，往往因冠词关系，使词组意义发生很大改变；b.介词的省译：英语中使用介词非常频繁，在某些名词、形容词、副词、动词后常有固定搭配的介词，这些搭配的介词，一般不译。在时间状语、地点状语，定语、状语、表语中的介词，也常省译。c.连接词的省译：英语中的连接词用来表示前后词语的逻辑关系和语法关系。汉语中把句子成分按逻辑关系排列后，就毋需用连接词来表示上下文的关系。有时加些虚词，如"就""却"等。d.某些动词的省译：在不影响完整表达原文内容的前提下，汉语译文常省译动词。这些动词一般是连系动词或作连系动词用的动词。e.形式主语或形式宾语"it"的省译。

这句"Turn the output on by pressing the OUTON button."里的介词"by"在译文中省译，"按下开启输出(OUT ON)按钮打开输出"。在下划线句子"Use the voltage knob and set the UVL level of the unit to 50% of its full-scale voltage rating or 30 volts, whichever is lower."，其中的代词"its"无须翻译，该句译为"…UVL电压设置为最大额定值电压的50%或30伏…"。在下划线句子的这个部分"... while the load current varies as required by the load."，无需把"while"翻译为"当……时候"，可以省译该词，也不会影响句子的意思表达。当然也可以把该词译为"此时"，可以达到一样的交际效果。

④ 翻译方法中的词汇转译法，就将原文中的某一个词由原来的词类转译成

目标语中另一词类，由于使用这种方法能够使词汇保持其基本概念意义（郭著章，2003）。a.名词、副词、形容词、介词转译为动词：现代英语中，名词的使用范围比较广，而汉语中动词使用得比较多。通常情况下，具有动作意义或含有动作意味的名词往往可以转译成汉语动词。英语多静态，而汉语多动态，一些表示状态的形容词常常会译成动词。表示知觉、情感、欲望等心理状态的形容词，在系动词之后作表语时，往往译成汉语的动词。汉语介词大部分是由动词虚化来的。英语介词在一些搭配中常起动词的作用，用来表达动作意义，汉译时，可将介词转译成动词。如在作表语的介词短语中，介词常转译为动词，而连系动词则省略不译。b.动词、形容词、代词转译为名词：名词化是英语的一个重要句法结构，是一种其他的词性（如动词、形容词、副词等）转换为名词的结构。来自动词的名词经常表示原来的动作或状态。英译汉时，这类短语大多还是译成动词。由形容词转化的名词（部分转化和全部转化两类）英译汉时常译成名词。有些代词需要做明晰化处理，此时最好是转译为名词。c.名词、副词转译为形容词：英语中直接作定语修饰名词的名词，有些作表语或宾语的抽象名词，以及某些形容词派生的名词，往往可转译成汉语中的形容词。另外，当英语动词转译成汉语名词时，原来修饰该动词的副词也往往随之转译成汉语中的形容词。

这句“Adjustment of the output current limit can be made when the output is enabled (ON) or disabled (OFF).”里面的“adjustment”是动词派生出来的，所以直接翻译为动词，还同时省译了介词“of”，“当输出开启(ON)或关闭(OFF)时，可以调整输出电流极限。”这句中的从句部分“…while the voltage varies as required by the load.”里面的“require”转译为名词后，还能够表示原来的动词意义，还同时省译了介词“by”，“此时电压随负载的不同需求而变化。”。这句“This should trip the OCP protection.”里的代词“this”本身所指的是上一句“Press the OCP/488 button”这个点击按钮的动作，所以为了明晰化的效果，代词转译为名词，翻译为“该动作会触发过流保护。”

翻译理论及应用讲评：术语学

科技术语的翻译不是机械的术语转换，而是要求译者首先有一定的术语理论知识,并在实践中根据具体文本语境进行选择。术语(term)是表达一定专业知识领域的词或短语。根据《新牛津英汉双解大词典》:a term is a word or phrase used to describe a thing or to express a concept, especially in a particular kind of language or branch of study.术语有科技术语和人文社科术语之分。本部分主要指的是科技术语。科技术语的典型特征是术语的单义性、系统性、准确性、简洁性、稳定性、理据性和多产性。其中术语的单义性和系统性是其最基本的两个特征。单义性是在一个学科领域内,一个术语只表达一个概念,同一个概念只用一个术语来表达。但在具体使用中,许多术语都会出现若干个变体形式,也会形成多义或者同音异义的现象。术语变异(variation)产生的原因在于不同民族对于科学概念的认知差异、语言差异和术语承载信息的不同等。系统性是在一个特定领域的各个科学术语,必须处于一个明确的层次结构中,共同构成一个概念系统,如属种关系(generic relations),整体——部分关系(partitive relations 或 meronymy),联想关系(associative relations),两极关系和分级关系(bipolar relations and gradeness relations),蕴含或者承袭关系(entailment relations)。

基于以上特征,科技术语翻译要以明确、简洁和规范为指导原则。在英语科技术语翻译为汉语时,要根据专业领域内的概念和语言结构语义准确术语的科学含义,确切地表达出其语义内容。科技术语的目的是把科技知识和信息传递给使用者,其语言形式要从科学概念体系出发,用精练的语言进行译述,要简单明了并且专业化。现有科技术语的译名要规范统一,要尽可能参照专业译法,通过查阅专业技术词典,尽可能避免译名创新,以免出现外行话。对于新科技术语,在了解相关领域专业知识后,基于对相关专业语境的理解,可使用音译法、意译法、形译法、零译法和构词分析法,尝试译出创新术语或者新出现的术语词义,并建议在译名后用括号附注原文。

科技术语形式多样,一般以三种形式存在,即单词式、复合式和短语式,比如“voltage 电压”“output voltage 输出电压”“turn-on check-out procedure 开机检查过程”。而术语的构成方法可分为派生法、复合法、形态语义法(包括缩略词,外部借词和内部借词)。派生词可按照构词法的规律,将科技术语分解为词缀和词根,解析出词素及其意义,然后调整词素意义,得出完整的词义,一般采用直译的方法转换为汉语。如例⑤“The green OUT ON indicator should be illuminated./ The green CV indicator should also be illuminated. 绿色的 OUT ON 指示灯应发亮。/绿色恒压(CV)指示灯也应该亮起。”中的

"indicator"解析为"indicate/指示 + or/动作的发出者",结合学科特点和术语语境,使用直译译为"指示灯"。英语词缀及词根大都来自拉丁语和希腊语,因此科技译者最好要熟悉一些此类的词缀和词根的意义。

复合词术语多采用音译、意译、音意结合、直译、形译的方法。所谓音译就是根据英语单词的发音译成读音与原词大致相同的汉字。用于音译的汉字一般只取其语音和书写形式,不再保留自身的原义。一部分音译一部分意译时,可能音译在前意译在后,也可能意译在前音译在后。而直译适合于大部分的复合词术语,即先解析出各合成语素的词义,然后直译成汉语词组或者短语。如例⑥"Sense connector pins 1 and 2 are jumped; sense connector pins 4 and 5 are jumped. /All switches on Connector J2 are in the down position. 传感器连接器接头引脚 1 和 2 断开;传感器连接器接头引脚 4 和 5 断开。/连接器 J2 上的所有开关处于断路位置。"中的"sense connector"解析为"sense/(传)感(器) + connect(or)/连接(器)",顺序直译即可。用英语字母加上名词构成的复合术语通常采用形译法。英语术语中英文字母表示某种事物的外形,具有隐喻色彩。翻译时选用近似这种字母形状的汉字词汇来表示其外形,或直接保留原字母来表示其外形,这就是形译。

外部借义词的翻译起初多采取音译的方法,但是音译不如意译能够明确表达术语的含义。因此有些起初音译词后来都被意译词取代,如"volt/伏特,电压"和"hertz/赫兹,频率"。此外,有时在音译时会在音译语素后增补表示类别的汉语语素,如"covar/科伐合金"和"logic/逻辑电路"。内部借义词多采取引申和借用的方法,即借用能够表达形状相似、功能相似、结构相似的汉语词语,如例⑦"Rotate the voltage knob while observing the DC VOLTS display. 旋转电压旋钮,观察直流电压(DC VOLTS)输出显示屏。"和"Press the OVP/UVL button once so that the DC AMPS display indicates OVP. 按下过压保护/欠压锁定(OVP/UVL)按钮一次,直流电流(DC AMPS)显示屏应显示 OVP (overvoltage protection)。"其中的"knob"原义是"球形突出物,瘤",译为"旋钮或球形把手",使用的是两者的结构类似。其中的"display"原义是"显示",译为"显示屏或显示器",增加的具体化的名词"屏/器",使用的是两者的功能类似。

缩略词一般用直译的办法,如 CV/CC(constant voltage/current)译为恒压/恒流,OVP/OCP(over-voltage/over-current protection)译为过压/过流保护。有些常用缩略词已经为大家所熟悉,此时这些常用福德英语科技缩略词可用零翻译法，直接引用此类缩略形式而不加翻译。有时英语缩略词和其他词一起构成短语类型的名词术语，译文可直接在英文缩略词后加其他的汉语译文，而不必将缩略词的含义译出。如例⑧"Press the OVP/UVL button twice so that the DC AMPS display indicates UUL. 按下 OVP/UVL 按钮两次，DC AMPS 显示屏应显示 UVL(undervoltage lockdown)电压值。"在本章英语文本的汉语译文中，缩略词第一次出现的位置采取了直译的方法全译出其含义，而在其后的再次出现，多采用了零翻译的形式。如有和其他词构成短语，缩略词第一次出现全译，再次出现也多采用零翻译加其他汉语译文的形式，不必每次都把缩略词代表的汉语意思译出。

补充阅读：Substation Equipment: Relay Room and Substation Configurations

1. Relay Room

The relay room contains the 'brains, of a substation. The relay room has three essential roles to ensure the substation is operated reliably and safely:

• Control. Substation control allows the operation of equipment within the substation by operators. If maintenance is required on a piece of equipment, an operator can switch out sections of the substation to make the equipment safe to work on, while maintaining power supply to consumers.

If there is a fault on a circuit connected to the substation, the protection will automatically disconnect the faulty equipment. An operator may then reconfigure the network to restore supply.

• Protection. The role of protection is to initiate the operation of a circuit breaker to clear a fault at or near the substation as quickly as possible to minimise damage to equipment, people, and loss of power to consumers.

• Communications. A substation has communications equipment installed so the status of the power flow through multiple substations can be observed from a central operating room. This allows the power system to be monitored as a whole; and plays a vital part in the dispatch of generation, and the allocation of sufficient reserve capacity on the power system. A Supervisory Control and

Data Acquisition System (SCADA) interface forms the main backbone of the power system communications system. Communication is also used to aid the operation of protection equipment. Communications are the systems and processes that are used to communicate or broadcast information, especially by means of electricity or radio waves.

As the power supply to the control room essential for managing the flow of power through the substation, there are back-up battery power supplies to maintain operation for a sufficient amount of time during a conventional power outage.

2. Substation Configurations

Different substation configurations are used for substations with differing levels of importance. For a 330 kV bus, some power company requires a minimum of a mesh configuration to provide some security against a bus fault occurrence in order to meet the reliability requirements of the Technical Rules. A mesh configuration is not readily expandable and so if the site is likely to have more than six switches, a breaker and a half configuration is used.

2.1 Single Busbar (Single Breaker Per Circuit)

The "single busbar" arrangement is simple to operate, places minimum reliance on signalling for satisfactory operation of protection, and facilitates the economical addition of future feeder bays.

A single busbar arrangement generally will have a circuit breaker present on each connected circuit, which will protect the substation from outage during an external fault. However, a bus fault will cause a complete outage of the substation until the cause of the fault can be resolved.

A single busbar arrangement can have extra security of supply by installing a "bus section breaker". A bus section breaker will split the bus into two sections, allowing half the bus to continue operating following a bus fault on either section.

2.2 Mesh

A substation with a mesh configuration has the same number of circuit breakers as the number of circuits. The scheme offers better features and facilities than the single busbar without a bus section breaker. However, it cannot easily be expanded in the event that additional circuits are required unless the initial design has been made suitable for future expansion. A mesh substation is usually limited to six circuits.

2.3 Breaker and a Half (1.5 Circuit Breaker Per Circuit)

A "breaker and a half configuration" uses three circuit breakers to connect two circuits (or 1.5 circuit breakers per circuit). This arrangement offers the benefits of a double busbar arrangement, as well as offering an increased level of security. The breaker and a half arrangement has the ability for each circuit to switch between busbars without an outage. Supply is maintained for a single busbar outage. Some level of supply is also maintained for the outage of both busbars depending upon the arrangement of the circuits.

This configuration provides the same level of reliability as the double busbar, double breaker arrangement, with fewer circuit breakers.

Chapter Eleven Positive and Negative Electrode Materials for Li-ion Technology

第十一章 锂离子电池正极和负极材料

原文

1 Positive Electrode Materials for Li-ion Technology

After 20 years of isolation, the research on Li-ion technology became highly diversified, especially with the arrival of nanomaterials, which opened up many opportunities. Technological breakthroughs are still to be made on electrode materials, electrolytes and formulation to improve their performances, reliability and safety. This is a broad area including different accumulator components.

The quest for high-performing electrode materials is now fulfilled by $LiMO_2$-type lamellar compounds and polyanionic compounds with the formula $AMXO_4$, with XO_4^{n-} being a polyanion, A an alkali and M a 3D metal. Their structure is an association of interconnected ① MO_6 octahedra and ① XO_4 tetrahedra within which are

译文

1 锂离子电池正极材料

锂离子电池技术经过20年的独立发展,其研究变得十分多样化。特别是随着纳米材料的出现,带来了很多机遇。除此之外,电极材料、电解质和配方材料的性能、可靠性和安全性改进还亟待技术突破。这是一个包含不同蓄电池组件的领域。

对于高性能电极材料的需求,目前已通过合成 $LiMO_2$ 型层状化合物和分子式为 $AMXO_4$ 的聚阴离子化合物实现,其中 XO_4^{n-} 为聚阴离子,A 为一种碱金属,M 为一种三维金属。它们的结构是相互连接的 MO_6 八面体型(M代表过渡金属)和 XO_4

cavities occupied by Li^+ ions. ⑤ This structure is not a very good conductor of lamellar compound because the MO_6 octahedra are isolated by XO_4^{n-} tetrahedra; these materials are, therefore, insulators and have been discarded as electrodes. With the transfer to nanometric scales, these materials became very attractive through the inductive effect of the polyanion with adjustable electronegativity, to adjust the redox potential $Mn^+ \rightarrow M(n-1)^+$ of the 3D element. Nowadays, studies focus on $LiFeBO_3$ and Li_2FeSiO_4 phases with a detailed understanding of the mechanisms involved in the cycling of batteries. Special efforts have been made to improve the electrode kinetics by either introducing conductive polymers such as additives around the particles, or by developing new positive composite electrodes by nanoconfinement of active polyanionic species (vs. Li) in mesoporous carbon.

四面体型(X代表金属元素)阴离子的组合,其内部是由Li^+离子占据的空腔。因为MO_6八面体被XO_4^{n-}四面体隔离,这种层状化合物不是良导体,所以这些材料为绝缘体,不能作为电极。随着纳米级材料的开发,这些材料变得非常有吸引力,通过利用具有可调节的电负性的聚阴离子诱导作用,可以调节三维金属元素氧化还原电势$Mn^+ \rightarrow M(n-1)^+$。目前,有关$LiFeBO_3$和$Li_2FeSiO_4$的研究集中在对电池循环过程中机制的研究,通过在颗粒周围引入诸如添加剂的导电聚合物,或者通过介孔碳上活性聚阴离子物质的纳米限制,来配合开发新的正极复合电极。

In the attempt to increase redox potential, the addition of a ②highly negative element such as fluorine has always been considered, such as fluorophosphates already mentioned by J. Barker and in particular phase $LiVPO_4F$ due to its high voltage (4.0 V). Through the production at low temperatures, a whole new family of fluorosulphates with the formula $AMSO_4F$ was developed by RS2E, such as the

为了增加氧化还原电势,经常考虑加入高电负性元素如"氟",还例如J.Baker已经提到的氟磷酸盐,尤其是有高电势(4.0V)的$LiVPO_4F$(氟磷酸钒锂)。通过低温生产,法国电化学能量储蓄RS2E网开发了一种全新的分子式为$AMSO_4F$的氟磷酸盐新族元素系列,如具有三维结构的$LiFeSO_4F$(氟硫酸亚铁

①LiFeSO$_4$F phase with a triplite structure, which inserts and disinserts lithium at a voltage of 3.9 V; this is still the highest redox potential for the couple Fe^{3+}/Fe^{2+} to have ever been reported for an inorganic compound. This sulfate chemistry has recently extended to cover hydroxy-sulfates, oxy-sulfates and even non-substituted $Li_2Fe(SO_4)_2$ sulfates, which have a redox potential of 3.83 V, and other pyrosulfates or even fluoropyrosulfates are now targeted. Whether sulfates or phosphates, these materials are also interesting since they are composed of abundant chemical elements and can be obtained from low-energy processes, thus obeying the concepts of sustainable development. ④However, their heavier weight, compared to that of oxides (due to their polyanion) stops them from competing with lamellar oxides in terms of performance, which makes them restricted to uses in large scale applications.

锂),可在一定电压下嵌入和脱出锂,其电势为3.9V,这是无机化合物中对 Fe^{3+}/Fe^{2+} 的最高氧化还原电位。这种硫酸盐类化学物质最近已发展到羟基硫酸盐,氧硫酸盐和甚至不可替代的硫酸亚铁锂[$Li_2Fe(SO_4)_2$],其氧化还原电势为3.83V。其他焦硫酸盐或甚至氟磷酸盐现在还有不足,无论是硫酸盐还是磷酸盐,这些材料都是有潜力的,因为它们是由许多化学组成元素丰富,并且可以从低能量化学反应中获得,从而符合可持续发展的概念。但是与氧化物(聚阴离子)相比,它们的重量更重,从而使它们无法在性能方面与层状氧化物竞争,这也就限制了它们的大规模应用。

Thus, RS2E also became interested in $LiCoO_2$ and Li-NMC (acronym for Ni, Mn and Co) lamellar compounds, currently used in smartphone batteries. It has recently been shown that similar compounds containing excess Li (called lithium-rich NMC) may allow capacities greater than 250 mAh/g, i.e. an increase in autonomy of our portable devices by

因此,RS2E也对 $LiCoO_2$ 和 Li-NMC(镍Ni锰Mn和钴Co的缩写)的层状化合物表现出兴趣,最近这项技术被用在智能手机的电池上。最近研究表明,一种类似的富锂离子化合物(称为富锂离子NMC),容量有可能超过250 mAh/g(毫安时/克),也就是在我们的便携设备当中,其续

more than 30%. However, due to major challenges, in particular the drop in mean voltage of the accumulator as a function of the number of charge and discharge cycles, these materials have still not been commercialized, despite extremely active and competitive international research.

航能力会提高 30%。然而，由于这种技术也面临着一些重大挑战，尤其是在充电和放电的循环过程中，存在着蓄电池的平均电压衰减。正是由于这些原因，尽管在国际上的研究十分活跃和富有竞争力，这种材料仍然未商业化。

To shed light on these problems, RS2E synthesized new families of lamellar oxides, ① $Li_2Ru_{1-x}M_xO_3$ (M = Mn, Sn, Ti, etc.) which are structurally similar to NMC but with a simpler chemical composition. By combining XPS and EPR spectroscopies, it was possible to reveal the reversible participation of oxygen in redox processes. This, therefore, proved that the accumulation of cationic and anionic activity within the same material was responsible for this increased ③ capacity. Moreover, out of the substituted compounds, only those substituted with tin showed a decrease in voltage. Comparative measurements using high-resolution microscopy on these materials substituted with titanium or tin revealed that this drop in voltage during cycling in these materials was due to the migration of cations and the capture of small cations (Ti) in the tetrahedra during cycling. This knowledge must now be applied to stabilize the voltage in Li-rich NMC phases, which is not at

为了解决这些问题，RS2E 合成新族层状氧化物 $Li_2Ru_{1-x}M_xO_3$（M_x = Mn 锰，Sn 锡，Ti 钛）锂离子电池正极材料，其具有和 NMC 相似的结构但是其化学构成相对简单。通过 X 射线光电子能谱分析（XPS）和电子顺磁共振光谱分析（EPR），就可以揭示氧的可逆参与氧化还原过程。因此，这也证明了活性阳离子和阴离子的大量集聚是电容提升的主要原因。除此之外，在被替换的化合物中，相同材料中只有那些被锡替换的物质表现出电势下降。通过运用高分辨率的电子显微镜可以观测发现到，运用钛元素和锡元素的替代物质对比，都揭示电压的下降是由于在循环过程之中，在四面体结构中徙动的阳离子和被捕获的钛元素小阳离子。这一项研究结果一定可以用在富 Li-NMC 阶段稳定电压，但是仍然不是很明显，以及为了探索这一新现象而

all obvious, and to better explore this new phenomenon to obtain materials with capacities far beyond what we could hope for five or six years ago.

获取所需要的新材料，让电池容量在未来五六年内达到我们所期望可能的样子。

2 Negative electrode materials for Li-ion technology

2 锂离子电池负极材料

With regard to negative electrodes, nanomaterials have certainly brought plenty of hope, but the solution is clearly not ideal if we consider the withdrawal of both SONY li-ion technology and PANASONIC silicon technology from the market, and no one knows for how long. The reason for this withdrawal is the same for both systems and is related to their limited performances during cycling especially at high temperatures. This is one of the reasons why negative electrode compounds comprising an alloy material that reacts with lithium have been studied by RS2E.

关于纳米负极材料的出现无疑带来了充足的希望，但如果我们考虑到索尼锂电池技术和松下硅技术从市场中撤出一段未知的时间，且无人知道会持续到什么时候，这一解决方案显然并不理想。两种技术退出的原因相同，都是与它们在循环中特别是在高温下的性能有限有关。这也是为什么 RS2E 已研究包含与锂反应的合金材料的负极混合物的原因之一。

They stand apart from the studies in the literature, especially those with Si, that attempt to create elegant "Yolk-shell" or "pomegranat", architectures to optimize the Si electrode that are totally unrealistic from an applicative standpoint. We prefer to focus our upcoming research on a chemistry aimed at (1) controlling the density and thickness of the SiO_x layer surrounding the silicon particle and (2) grafting elastic or even conducting polymers.

RS2E 的研究与诸多文献研究不同，尤其是那些有关硅的文献，其中有些试图构架优雅的"蛋黄壳"或"石榴"结构。从适用性的角度看，这些优化 Si 电极的构想完全不现实。我们倾向于将我们即将开展的化学研究集中在(1)控制硅颗粒周围 SiO_x 层的密度和厚度，(2)移植弹性结构或甚至导电聚合物。对其他一些与锂进行合成的有

④ The design of other promising compounds that can alloy with Li such as Li_xTiSnSb has also been studied. Finally, we note that through the confinement of the carbon matrix, high-performance SnO_2-based electrodes could be created, despite the problem of irreversibility which is still preventing the use of such electrodes.

开发前途的化合物，例如Li_xTiSnSb(锂钛锡锑合金)等也开展了研究。最后，我们注意到，通过碳基体约束，可以产生高性能的以二氧化锡为基础的电极。但不可逆性问题仍然阻碍这种电极的使用。

《术语表》

Li-ion	锂离子
electrode material	电极材料
lamellar compound	层状化合物
polyanionic compound	聚阴离子化合物
redox potential	氧化还原电势
mesoporous carbon	介孔碳
fluorine	氟
fluorosulphate	氟磷酸盐
pyrosulfate	焦硫酸盐
accumulator	蓄电池
RS2E	法国电化学能量储蓄网
NMC	镍 Ni 锰 Mn 和钴 Co 的缩写
XPS	X 射线光电子能朴分析
EPR	电子顺磁共振光谱分析
cationic and anionic	阳离子和阴离子
conducting polymer	导电聚合物
carbon matrix	碳基体

翻译难点讲评

① 化学分子式的翻译在原文多处出现，其翻译一般可采取三种方式，根据不同的情况选用合适的方法。一般化学分子式所表达的有机化合物英文名不出现在原文中时，并且分子式有着各自固定的英文名和译文名称时可

以采用零翻译。另外，如果当含有变量的化学式不是表示一种材料，而是单纯的一种化合物时，可保留原化学式的形式。如："Their structure is an association of interconnected MO_6 octahedra and XO_4 tetrahedra within which are cavities occupied by Li^+ ions. 它们的结构是相互连接的 MO_6 八面体型（M 代表过渡金属）和 XO_4 四面体型（X 代表金属元素）阴离子单元的组合，其内部是由 Li^+ 离子占据的空腔。"或者在保留化学式的同时，可以增加其汉语译文，放在括号中加注说明。如："... a whole new family of fluorosulphates with the formula $AMSO_4F$ was developed by RS2E, such as the $LiFeSO_4F$ phase with a triplite structure... 法国电化学能量储蓄网（RS2E）开发了一种全新的分子式为 $AMSO_4F$ 的氟磷酸盐新族元素系列，如具有三维结构的 $LiFeSO_4F$（氟硫酸亚铁锂）"。本例中第二个化学分子式就采用了分子式加汉语译文注解的方式翻译。但是也有些情况下若采取零翻译，不能完全符合汉语表达习惯。比如："To shed light on these problems, RS2E synthesized new families of lamellar oxides, ①$Li_2Ru_{1-x}M_xO_3$ (M = Mn, Sn, Ti, etc.) which are structurally similar to NMC but with a simpler chemical composition. 为了解决这些问题，RS2E 合成新族层状氧化物 $Li_2Ru_{1-x}M_xO_3$（M_x = Mn 锰，Sn 锡，Ti 钛）锂离子电池正极材料，其具有和 NMC 相似的结构但是其化学构成相对简单。"此例中的化学分子式中"M"是一个变量，它代表金属元素"锰，锡，钛"多族氧化物，所以前文所说这类材料都是锂离子电池正极材料，因此进行表达上的增译，以让化学式所代表的材料是多种化合物，并属于电极材料的属性明晰化。

② 本句中" In the attempt to increase redox potential, the addition of a highly negative element such as fluorine has always been considered..."。"element"在化学中表示"元素"，"negative"表示"负性的，阴性的"。故将"highly negative element"初译为"高负性元素"和"高阴性元素"。但验证电化学中是否存在该名词词组时，在术语在线（termonline. cn）上得到的结果是查无此词。但是查找到有关"电负性"（electronegativity）的术语，根据本句前文已经提及"为设法增加氧化还原电势"（In the attempt to increase redox potential）故将"highly negative element"译为"高电负性元素"符合原文语境，整句译为：为了增加氧化还原电势，经常考虑加入高电负性元素如"氟"。

③ 这个句子里的 capacity 是个多义词，其意思有"容量，能力，生产力"等，在

电化学中常用的词义有“容量”(指电池在一定条件下的放电能力),“电容”(给定电位差下的电荷存储的能力),“比电容量”(单位电池或活性物质放出的电量)。而根据该词所在句子上下文,“This, therefore, proved that the accumulation of cationic and anionic activity within the same material was responsible for this increased capacity.”提到了带电离子的集聚。查百度百科有如下信息:“电容亦称作‘电容量’,是指在给定电位差下的电荷储藏量,记为C,国际单位是法拉(F)。一般来说,电荷在电场中会受力而移动,当导体之间有了介质,则阻碍了电荷移动而使得电荷累积在导体上,造成电荷的累积储存,储存的电荷量则称为电容。”并且本段中所提到的新族层状氧化物 $Li_2Ru_{1-x}M_xO_3$ (M_x = Mn 锰,Sn 锡,Ti 钛)是锂离子电池正极材料。因为这些物质本身并不是电池,也不是活性放电物质,所以不可能是“容量”和“比电容”的含义。而“电容”是指电荷的存储能力,可用来说明电池材料在一定条件下所具有性质,符合该句语境,整句译为“因此,这也证明了活性阳离子和阴离子的大量集聚是电容提升的主要原因。”

④ 长句特点之一是含有较多的从句,比如定语从句,其中有限制性定语从句,也有非限制性定语从句,在原文本中,也不乏两种定语从句同时使用。由于定语从句具有延展性,即理论上具有无限叠加使用的特点,因此,科技文本中会较多采用这种表述方式,以达到对某些原理、现象或客观事实的阐述。但汉语则不同,并无定语从句这种句式结构。因此,译者要着重分析句子结构,准确定位修饰成分与先行词的关系,采取前置修饰语、后置独立单句、转换为其他类型的从句或者句子成分的方法。比如4中下划线的两个句子就分别采用了不同的翻译方法。在句子“... promising compounds that can alloy with Li...”里面,定语从句所修饰的先行词是“compounds”这个单词,先行词本身有一个形容词作前置定语,两个定语的内容和结构都不复杂,可以直接把定语从句前置,与原来的“promising”一起作定语,“与锂进行合成的有开发前途的化合物”。另外一个句子“However, their heavier weight, ... in terms of performance, which makes them restricted to uses in large scale applications.”中的定语从句修饰的是整个句子,只能考虑分译为一个单独的分句,“但是……,它们的重量更重,……在性能方面……,这也就限制了它们的大规模应用。”

翻译理论及应用讲评:关联翻译理论

关联理论(relevance theory)是丹·斯珀伯(Dan Sperber)和迪尔德丽·威尔逊(Deirdre Wilson)1986 年在专著《关联:交际与认知》(*Relevance: Communication and cognition*)中提出并发展出来的。他们的学生恩斯特·奥古斯特·格特(Ernst August Gutt)用关联理论对翻译进行研究,与 1991 年在专著《翻译与关联:认知与语境》(*Translation and Relevance: Cognition and Context*)中提出了关联翻译理论。关联理论是一个关于语言交际的理论,认为明示和推理是交际行为的两个方面,语言交际过程就是一个明示——推理过程(ostensive-inferential process)。其核心为关联原则,即每个明示交际行为都有最佳关联性,也就是话语理解时以最小的认知努力(process efforts)获得足够的语境效果(contextual effects)。所以交际者总是以明示手段提供最大的关联性信息,而交际对象总以推理手段选择关联度最大的假设,进而推导交际者的意图。关联翻译理论认为翻译也是一种推理过程,是一个寻找关联链和最佳关联的认知过程。关联性取决于两个因素,即处理努力和语境效果。同等条件下,处理努力越小,关联性越强;语境效果越大,关联性越强。译文就是要达到最佳关联性的目标。译者就是努力让原文作者的意图(intention)与译文读者的期待(expectation)平衡或者一致。为此,译者首先必须从原文各种交际线索中(communicative clues)通过解码和推理两个过程,综合解析出原文的意图或者内涵。同时,译者还需要考虑读者的认知环境,即包括了译文所传递的语境假设(context assumption)和读者能力(competence)之间存在的潜在语境差异(potential context),对译文进行取舍,最终找到能够达成原文作者交际内涵和读者认知期待最佳关联的译法。译文应该是同原文释义相似(interpretively resembling)的目标语语段。

根据关联翻译理论,科技翻译实质上也会"死"在最佳关联原则指导下的,对原语进行的明示推理交际行为。科技翻译中的各种交际线索来自译者的科技知识、语言知识和语境信息。下面以本章文本和译文为例说明关联翻译理论的应用。

例⑤原文"This structure is not a very good conductor of lamellar compound because the MO_6 octahedra are isolated by XO_4^{n-} tetrahedra; these materials are, therefore, insulators and have been discarded as electrodes."

译文为“因为 MO_6 八面体被 XO_4^{n-} 四面体隔离，这种层状化合物不是良导体，所以这些材料为绝缘体，不能作为电极。”这句所说的“This structure is not a very good conductor...”让人费解，因为导体和结构并不存在合理的关联。另外，原文中的“a very good conductor of lamellar compound”如果按照原文形式直译就会是“层状化合物的良导体”，明显存在专业逻辑错误。所以译者推理作者可能是把介词短语“of”的位置搞错了，实际应该位于“structure”之后，应该是“This structure of lamellar compound is not a very good conductor because...”。此句后半部分句子的主语对象是“these materials”，陈述说这些材料在电极材料中已经被弃用（discarded）。依据前文所说这种层状结构化合物不是良导体，推理作者可能是说这种材料其实根本就不能用作电极材料。专业知识告诉我们，绝缘体是绝对不可作为电极使用的。因此原文中的“discarded”不能译为“放弃、抛弃、摈弃”之类的表达。因为这些译文都含有曾经用过它作电极材料，再转而不用的意思。此例可以显示出，作为交际线索的科技知识、语言知识和语境信息是为推理的前提和基础。译者是否具备一定的推理能力（inferring ability），决定了科技翻译的有效度（transaltion validity），即是否成功实现了交际目的以及科技翻译的完美度（translation perfectibility），是否达到了原文与译文内容的无限接近。

补充阅读：Fault

A fault on a power system is caused by a breakdown of insulation between the main electrical conductor (or conductors) causing an interruption to normal power flow. These events are unpredictable and unavoidable. In order to avoid damage to the remainder of the network, and (further) disruption to users, the faulted components are disconnected by ‘protection systems’

Factors that may cause the insulation to breakdown include:

- atmospheric conditions, such as lightning, wind etc.
- pollution forming a current leakage path across exposed insulators
- vermin
- equipment failure
- human error (incorrect switching)

1. Types of Faults

There are two types of faults: "balanced faults" and "unbalanced faults".

• Balanced fault. A three-phase fault involves the short-circuit of all three phases of one electrical circuit. A three-phase fault is also referred to as being a balanced fault as all fault current flows through the conductors without flowing through ground; and all three phases are affected evenly by the fault. A three-phase fault on a system generally causes the greatest amount of disruption to the power system. This is because the short-circuit causes the voltage and corresponding power transfer capability to collapse on all three phases.

• Unbalanced fault. Unbalanced faults are most common and are usually caused by "single-phase to ground" faults. An unbalanced fault describes any fault that causes fault current to flow unevenly in the three phases of the faulted circuit. The most common kind of unbalanced fault is a "single-phase to ground" fault. This is especially true on transmission lines, where the likelihood of short-circuiting between two phases is low due to a large separation between phases.

2. Stages of a Fault

A single-phase to ground fault within a terminal substation will have higher fault current than a three-phase fault; however, will generally cause less disruption to the overall power transfer capability as the two 'healthy' phases are able to transfer a limited amount of power.

• Pre-fault. The "pre-fault" conditions of the power system largely determine the amount of current that will flow, should a fault occur.

• Sub-transient. The "sub-transient stage" deals with the initial surge of current that flows into the fault once a fault has been established. The current peak within this stage determines the mechanical shock loading on equipment. The energy contained in the current surge determines how quickly equipment will overheat.

• Transient. The "transient stage" looks at the tail end of the current surge that occurs after the fault inception. Circuit breakers are often required to break the current flow during this stage to isolate the fault and help the system recover. However, technology advances made in circuit breakers and relays are allowing more responsive action, and may require the circuit breaker to be rated to breaking current at sub-transient levels.

• Steady state. The "steady state fault current" is an indication of the current that would flow if the fault was left uncleared on the system, after the initial surge has stabilized. The steady state of a fault is rarely observed on transmission circuits as a fault is generally cleared during the initial current surge. However, in distribution circuits, a remote fault on the system is more likely to appear similar to a large load. The fault may take longer to clear, therefore approaching the steady state current value.

3. Factors Influencing Fault Currents

The total current that flows at the point of a fault is made up of the contributions from many sources on an electrical network. The current contributed does not come solely from generators; motors may also contribute current to the fault that comes from the energy stored in the motor's rotating magnetic field.

Generally, a fault that occurs close to generation will have a high fault current, while a fault on a network point that is distant to generation will have low fault current.

• Impedance. The impedance of the lines between the fault point and the points of generation on the network will largely determine the value of fault current that will flow at the fault. The total impedance from the fault to the points of supply is known as the "source impedance" (or sometimes called "system impedance").

If the system impedance at a particular network point is small, then the point will have a large fault current, should a fault occur. The size of impedance is dependant on the number of circuits to which the point is connected and the voltage level. A point with small power transfer capacity will have high system impedance. The initial peak fault current is also affected by the ratio between the resistance and reactance of the source impedance. If the system reactance is significantly larger than the system resistance, then the initial DC current flow will be sustained for a longer period of time.

• Network strength. A network point that would have a large fault current flowing if there were to be a fault, is said to have a "high fault level".

Conversely a network point with a small fault current can be referred to as a point with a "low fault level".

The fault level of a network point determines the "strength" of the point, or the ability of the point to transfer power in a stable condition.

4. Fault Ratings

From this it can be said that a new generator connected to a strong point in the network (point with a high fault level) will have less trouble exporting power, from a stability point of view, than the connection of the generator to a weak point (low fault level) in the network.

In order to limit the scale of damage from a fault, all unfaulted equipment needs to be able to handle current flowing to the fault location until the fault can be cleared by protection. If this were not the case one fault would cause cascading failures and damage to the surrounding network.

The addition of new generation to an existing network may cause the fault level to increase beyond the fault ratings of equipment currently connected at this point. In this case equipment may need to be replaced with equipment of higher fault rating, or a device that reduces the fault contribution from the generator.

Chapter Twelve Some New Technological Information on Energy and Power

第十二章 能源电力科技发展新信息

原文

1 Tiny Graphene Sheets Make for Big Energy Storage

If you've ever used a smartphone and complained about the battery life, you've experienced one of the limiting factors on gadget design. Supercapacitors, which store charge rather than generate current with a chemical reaction the way batteries do, might offer a solution, but they tend to be bulkier than batteries and store less energy in the same size.

⑦But now a team of scientists thinks they may have found a way to create a compact, graphene-based super capacitor that lasts as long as a battery. Graphene, which is a single layer of carbon molecules, could lead to lightweight supercapacitors for electric cars and supplement traditional batteries both in electronic devices and renewable power systems.

译文

1 有助于大型储能的微型石墨烯薄片

如果你曾经使用过智能手机并且抱怨过电池续航能力不足的问题，那么你已经体会到了这种手机设计的受限之处。超级电容器也许是一个解决方案，与电池通过化学反应产生电流不同，它可以存储电能，但是它比电池体积大并且单位储能更少。

现在一个科研团队已经找到了一种以石墨烯为基础，但寿命可以媲美电池的紧凑型超级电容器的制作方法。石墨烯是一种由碳原子构成的单层片状结构的材料，它可以制作轻质超级电容，并可辅助应用于电子设备、可再生能源系统以及电动汽车等领域。这项由澳大利亚莫

The study, led by Dan Li of Australia's Monash University, is detailed in the Aug 1st issue of *Science*.

At its basic structure, a classic capacitor is comprised of two plates, usually metal, separated by some other material, sometimes a solution or simply air. When a charge is applied, electrons gathers on the surface of the plates but are limited by the plate's total surface area. ②Add more plates, and get more charge capacity.

Li's team made their version of a supercapacitor by folding sheets of graphene into corrugated shapes. Then they put the sheets into an electrolyte solution that kept the layers separated by a few nanometers and provided a path for moving electrons. Because of graphene's unique honeycomb lattice structure and because it was corrugated, there was more surface area for the electrons to gather upon, and thus the super capacitor could hold more charge in a smaller space.

In fact, more electrons packed together than anyone had managed before. Supercapacitors are typically made from porous graphite, and can hold about 5 Watt-hours per liter. Li's team managed an energy density of about 60 Watt-hours per liter, or about the same as a car battery. ⑧That isn't as good as lithium-ion batteries, which range anywhere between

纳什大学李丹教授领导的相关研究发表在2013年8月1日的《科学》杂志上。

一个传统电容器的基本结构通常由两个金属极板组成,其中灌满了电解质,有时是液体电解质,有时仅仅是空气电解质。当施加电荷时,电子在金属极板表面聚集却受限于板块总表面积。若金属极板的表面积增加,电容器的充电容量就会增加。

李丹教授的研究团队通过将石墨烯折叠成波浪形状制成了超级电容器。然后他们把这些薄片放入电解质溶液中,通过让这些薄片之间保持几纳米的距离为移动电子提供通道。由于石墨烯独特的蜂窝状晶格结构再加上波浪状,增大了电子聚集表面积,因此超级电容器就能在更小的空间储存更多电荷。

事实上,更多的电子前所未有的聚集在了一起。目前的超级电容器通常由多孔石墨制成,其能量密度为每升(体积)可容纳5瓦特时。李丹团队所做的超级电容器能达到每升(体积)60瓦特时的能量密度,与汽车电池相当。它虽然并不像锂离子电池能量密度那么高(锂离子

250 and 750 Watt-hours, but it's still a big jump over current capacitor technology.

电池能量密度在 250～750 瓦特时/升之间),但这在电容器科技上仍是一个巨大的飞跃。

2 Development of Cognitive Radio for Smart Grid

The current centrally controlled power grid is undergoing a drastic change in order to deal with increasingly diversified challenges, such as rising energy demand, aging infrastructure, renewable energy sources, reliability and security. These challenges have become a global concern that necessitates the traditional electrical grid to evolve toward intelligence, autonomy, improved efficiency, easy control, and high security. The next-generation power grid, known as the smart grid, will be realized with proactive usage of state-of-the-art technologies in the areas of sensing, communications, control, computing, and information technology. In a smart power grid, an efficiency and reliable communication architecture plays a crucial role in improving efficiency, sustainability, and stability. Optical fiber communication plays a very important role in the construction of smart grid because of small interference, large capacity and other advantages. However, it is unrealistic to achieve information connectivity wherever only rely on optical fiber communication. In this context, wireless

2 智能电网认知无线电技术的发展

目前,为了应对日益多样化的挑战,如能源需求的上升、基础设施老化、可再生能源发展的要求以及电网可靠性和安全性的要求等,集中控制中央电网正在发生巨大变化。这些挑战已经成为全球关注的问题,需要传统的电网向智能化、自动化、提高效能、易控制和高安全性发展。新一代电网即智能电网,将会积极应用传感、通信、控制、计算和信息技术等领域最先进的技术。在智能电网中,高效可靠的通信结构在提高效率、可持续性和维持稳定性方面起着至关重要的作用。光纤通信由于其干扰小、容量大等优点,在智能电网建设中起着非常重要的作用。然而,如果只依靠光纤通信,实现信息的连通性是不可行的。在这种背景下,无线通信技术得到了应用的机会。但是,无线通信技术受限于频谱资源短缺,即无线频段很少为电网专用。

communication technology gets the opportunity to application. However, the development of wireless communication technology is limited due to the short of spectrum resources, that few radio frequency band is power grid dedicated.

①Cognitive radio (CR) is an intelligent wireless communication system, it can perceive the external environment and learn from the environment by using of artificial intelligence. ③ Through real-time changes in certain operating parameters (transmit power, carrier frequency and modulation techniques, etc), the internal state changes to adapt to the statistical properties of the received wireless signal, to achieve any time, any place of a highly reliable communication, the spectrum resources to achieve effective utilization. Application of CR for smart grid will effectively address the problem of spectrum resource scarcity in the wireless communications. ⑤ A CR network can serve as a robust and efficient communications infrastructure that can solve both the current and future energy management needs of the smart grid.

认知无线电是一种智能无线通信系统,它可以通过人工智能感知外部环境,并从环境中学习。通过实时改变某些工作参数(如发射功率、载波频率和调制技术等),使得无线电内部状态发生改变,以便适应接收到的无线信号的统计特性,从而实现任何时间、任何地点的高度可靠的通信,和频谱资源实现有效利用。智能电网认知无线电的应用将有效地解决无线通信中频谱资源稀缺的问题。认知无线电网络可以作为一个强大而高效的通信基础设施,可以解决智能电网当前和未来的能源管理需求。

3 Unmanned Aerial Vehicles for Transmission Lines

3 输电线路无人机

⑥Unmanned Aerial Vehicle means a power driven aircraft, other than a model aircraft, that performs a useful mission

无人驾驶飞行器是一种除了模型飞机之外,执行有用的任务并且可以远程控制或具有自

and can be remotely controlled or has autonomous flight capacity. The term UAV has recently been expanded to Unmanned Aircraft Systems (UAS) and includes other elements such as communications to ground stations. This system combines Information and Communications Technology (ICT) flight management systems, on board sensors, synthetic vision systems and radar with ground based database systems and human analysis. Using UAS for transmission applications has great potential for both routine and emergency line inspections. In addition, recording digital audio and video observations could provide time stamped and geospatially accurate inspection data. Access to real time and interactive data is a key feature of the smart grid concept. The capacity to collect reliable data at low cost using UAS and ICT will enable the support and operation of a smart grid. The applications of unmanned aerial vehicles for electric transmission lines are as follows: Damage assessments application; Planning and forecasting work; Vegetation control work for transmission lines with UAS; ④ Right-of-way access management work; Structural inspection and condition assessment; Fault and damage assessment for line reclose; Line security monitoring. To achieve these applications, the various requirements were synthesized to prepare as follows:

主飞行能力的动力飞行器。无人机这个术语最近已经扩展到无人飞行器系统(UAS),包括其他因素,例如与地面站的通信。该系统将信息和通信技术(ICT)、飞行管理系统、车载传感器、合成视觉系统、配备地面数据库系统和人工分析的雷达相结合。将无人飞行器系统应用于输电,对于常规和紧急线路检查具有很大的潜力。此外,记录数字音频和视频观测情况可以提供时间戳和地理空间精确的检查数据。实时和交互式数据的访问是智能电网概念的一个关键特征。使用无人飞行器系统和信息通信技术以低成本收集可靠数据的能力将使智能电网的支持和运营成为可能。输电线路无人机的应用如下:损坏评估应用;规划和预测工作;无人飞行器系统输电线路植被防治工作;道路通行权管理工作;结构检查和状态评估;线路重合的故障和损坏评估;线路安全监控。为了实现这些应用,各种性能要求综合如下:

UAS has to have real-time data streaming capability to ground based control station; Video and still images collected by UAVs should be full high definition (HD);

无人飞行器系统必须对地面控制站具有实时数据流功能；无人机收集的视频和静止图像应该是全高清(HD)的；

UAV should have GPS capability; UAV should have automated detection and collision avoidance in the event of a control data link loss; UAV should be highly maneuverable so that it can operate in close proximity with transmission lines over very steep and mountainous terrain;

无人机应具有 GPS 功能；在控制数据链路丢失的情况下，无人机应该具有自动检测和防撞功能；无人机还应该具有高度的机动性，以便它能在非常陡峭和多山的地形上靠近输电线路运行；

UAV should be capable to operate in 60km/h wind, also in rain, snow and sleet;

无人机应该能够在 60 km/h 的风速下行驶，也可以在雨、雪和冰雹中行驶；

UVA control systems should be visible or configurable to air traffic controllers;

无人机控制系统对于空中交通管制员应可视或可配置；

UAV should be considered to have the same flight safety requirements as manned vehicles;

无人机应与载人飞行器具有相同的飞行安全要求；

UAV should be easily controllable with semi-autonomous and autonomous flight operation modes;

无人机应该易于使用半自主和自主飞行操作模式进行控制；

UAS should be capable to operate in the Electro Magnetic Fields (EMF) around transmission lines up to 500k V;

无人机应能够在 500kV 以上传输线路周围的电磁场(EMF)中运行；

UAV should have hovering capability and stability and be able to take off and land in a very small area;

无人机应具有悬停能力和稳定性，能够在非常小的区域内起飞和着陆；

UAV size should enable transportation in a regular sized work pickup truck and have a portable field ready control station;

无人机的尺寸应该能够在一个普通大小的工作皮卡车上进行运输，并有一个便携式现场准备控制站；

UAV should be able to airborne at least for 2 hours and travel at least 15 km;

无人机应该至少能够在空中飞行 2 小时，行驶 15 公里；无人机应该能够飞出视线之外；

UAV should be capable of flying out of line of sight; UAV should have a robust transponder system or similar to enable accident recovery;

无人机应该有一个强大的应答器系统或类似的设备，以实现事故恢复；

UAV should have a Maximum Take-off Mass (MTOM) not exceeding 150 kg.

无人机的最大起飞质量(MTOM)不得超过 150 千克。

There are many UAV technologies developed and built for various purposes.

目前有许多为各种目的而开发和制造的无人机技术。

The most effective, reliable and well known UAS are those developed for military and defense applications. The major military systems largely take advantage of recent developments and improvements to GPS navigation, image recognition and processing, new materials and ICT.

最有效、最可靠和最知名的无人飞行器系统是为军事和国防应用开发的。主要的军事系统很大程度上利用了 GPS 导航、图像识别和处理、新材料和信息通信技术方面的最新发展和改进。

《术语表》

Graphene	石墨烯
supercapacitor	超级电容器
electrolyte solution	电解质溶液
optical fiber	光纤
smart grid	智能电网
Cognitive radio	认知无线电
Unmanned Air Vehicle	无人驾驶飞行器

翻译难点讲评

① 对于此句里出现的术语“cognitive radio”到底是翻译成为“认知无线电”或者是“感知无线电”，译者本着科技词语词义选择的专业化原则和搭配原

则进行了分析。首先知网翻译助手对于“认知无线电”和“感知无线电”的使用频率查询结果是,前者 26 次,后者 5 次。比如前者使用的上下文例句有“OFDM technology will become the core technology for wireless communication in the future,OFDM will be widely used in cognition radio,in the 4G mobile communication and the wireless broadband access. OFDM 技术将成为未来无线通信中的核心技术,在认知无线电、4G 移动通信和无线宽带接入等领域得到广泛的应用(《信息安全与通信保密》2007 年 05 期)”。而后者出现的例句有“Cognitive radio has both the ability to sense the spectral environment and the flexibility to adapt transmission parameters. 感知无线电可以自动感知外界的无线环境并能调整系统参数以适应环境的变化(《通信技术》2007 年 05 期)”。比较有趣的发现是两者都主要同时出现在大量的北京邮电大学的硕博论文中。到底哪个才是最佳的译文？百度百科对感知和认知的解释是这样的:“感知是和感觉更密切,认知主要过程包括的就有感觉和知觉,两者一般不能分开,感觉是浅层次的,知觉是深层次的。感知仅仅是通过感觉到知觉的一个简单的加工过程,不涉及大量的抽象思维和本质的认识。认知则包括感知,认知涵盖的范围大。我们的所有知识的获得即依靠认知能力,其中有感觉、知觉、记忆、思维等方面。”简而言之,感知通过感觉器官与外界的作用而知道,而认知是更为深层次的知道,是通过思考,有意识参与的情况下达到的认识。由此可以看出,认知肯定比感知的包括范围广,是上义词。最终选择“认知无线电”,并且该用法还出现在《信息工程大学学报》2011 年第 3 期的国家科技重大专项资助项目成果论文中,双语平行文本为:“Spectrum sensing is one of the key technologies in cognitive radio, and cooperative spectrum sensing can improve the spectrum sensing performance. 频谱感知是认知无线电系统的关键技术之一,协作频谱感知能够提高频谱感知性能。”

② 科技英语有很多特点,而某些特点有时对于理解能产生很大的困难。广泛使用省略结构就是这种特点之一。它常常被我们所忽视,但确实又是一个值得注意的问题。科技英语中的省略现象分为两类:结构省略和上下文省略。在两者之中,重点是较为常见的上下文省略。例如这里句子“Add more plates, and get more charge capacity.”就存在上下文省略。根据该段所述内容,主要讨论的是“传统电容器的两块金属极板,受限于总表面积,而充电容量不足”。所以,该句译为“若金属板块的表面积增加,

电容器的充电容量就会增加。”

③ 译者应当从原文上下文的关系中，来探求正确的译法，只有从上下文关系中透彻地理解了原文的真意，才有可能写出准确、通顺的科技译文来。从读懂全文而后译、充分运用词法句法知识提高理解力、立足于所译的专业把握住句义和词义、防止望文生义。比如说此句，“Through real-time changes in certain operating parameters (transmit power, carrier frequency and modulation techniques, etc), the internal state changes to adapt to the statistical properties of the received wireless signal, to achieve any time, any place of a highly reliable communication, the spectrum resources to achieve effective utilization.”原句并不是一个完全符合语法的句子表达，可能缺少了主句。比如介词“through”的宾语包不包括后面的“internal state changes”。如果不包括，那么“changes”是动词还是名词词性？“to adapt”和两个“to achieve”是并列的么？这些问题都影响到句子的理解。从前文所说“认知无线电是一种智能无线通信系统，它可以通过人工智能感知外部环境，并从环境中学习”，以及后面所述“智能电网中的认知无线电的应用是为了解决无线通信中频谱资源稀缺的问题”。所以，本句可以译为“通过实时改变某些工作参数(如发射功率，载波频率和调制技术等)，使得无线电内部状态发生改变，以便适应接收到的无线信号的统计特性，从而实现任何时间、任何地点的高度可靠的通信，和频谱资源实现有效利用。”译文中增译了多处逻辑连接词，以使得原句中暗含的逻辑关系显化，比如“使得……以便……从而……和”。

④ 正确理解原文，包括正确理解词与句等语法现象、上下文所述事物发展的逻辑关系以及原文所涉及的具体事物本身等。忠实于原文，并不是不分青红皂白地、原封不动地译出了事，需要忠实的是原作者的真实意图。比如说原文中这处表达“right-of-way access management work”。查百度百科有以下解释：“路权最起初是用来形容出行不受阻碍，自由进入某段路径，而无须考虑土地所有权或其他法规一个术语。公共路权不受土地所有权限制，且允许所有人自由旅游与访问。在很长一段时间里，授予市民某段常用道路或轨道的路权是一种惯例。如果一片土地为个人所有且这片土地与其他人所有的土地接壤，则法院必须承认由这片孤立土地所有者核准的路权。”因此，“right-of-way”可译为“路权”。“access”的意思较多，有“通行权，使用权，访问权……”。鉴于两者的同类搭配：“transmission rights-of-way, controlled access rights-of-way",可得出正确的理解,译为"道路通行权"。

翻译理论及应用讲评:语义翻译和交际翻译理论

语义翻译(semantic translation)和交际翻译(communicative translation)是彼得·纽马克(Peter Newmark)在其1981年的专著《翻译问题探索》(*Approachesto Translation*)中提出来的。语义翻译的意思是译者只在目的语语义和句法结构的限制内,尽可能再现原作者的语境意义。与直译相比,语义翻译更尊重文本的语境。与忠实翻译相比,语义翻译相对更多考虑原文的审美价值,有时可采取折中手段只译出部分意义。交际翻译的意思是译作尽可能在目的语中再现原文读者感受的阅读效果。所以交际翻译以目的语读者为中心,尽力产生更有可读性的译文。而语义翻译侧重于原文译文形式的对等,更多保留原语文化、词汇、句法和语义,忠实于原文作者。语义翻译产生的译文倾向于比较复杂、难懂,注重细节信息集中和意义的细微差别,往往译文词汇增添了原文没有的意思,多属于超额翻译(over-translation)。交际翻译产生的译文更加流畅、简洁、清晰、直接,忠于目的语规范,倾向于在文字细节上对原文的调整、变更,多属于欠额翻译(under-translation)。

在纽马克的翻译理论体系中,语义翻译和交际翻译是有区别的,但又有重合,有时两种方法可产生类似的效果,有时原文一部分适合采取语义翻译,另一部分适合采取交际翻译。二者互为补充,可以看作一个整体的两个不同方面,在使用中不要孤立起来。当然,也不能说一个就比另一个翻译效果好,而是要在翻译中配合着交替使用。

在1994年,鉴于语义翻译和交际翻译不能完全解决翻译过程中的问题,作为两者的补充,纽马克在《论翻译》(*About Translation*)中又提出了"关联翻译法"的新概念,即原作或者译出语文本的语言功能越重要,就越要贴近原文翻译。这里的"重要"取决于译文文本类型和使用场合以及原语的语言功能,比如某类文本是否综合了或者偏向于下列某种使用功能:表情功能(expressive function),信息功能(informative function),互换功能(vocative function),审美功能(aesthetic function),寒暄功能(phatic function),元语言功能(metalanguage function)。这里的"贴近"是既要保存词的概念涵义,也保存词的内涵含义,不但保存原词的意义,还保存其在修辞、音韵上的用法。

贴近除了词汇贴近外，还包括语法和句法贴近，即根据贴近程度选择复制原语言语法、句法或者对它们进行置换等操作。最高层次的贴近是文本贴近，是词汇加语法加句法贴近，再加上文本衔接手段和连贯性的贴近，以及文本风格和语域（语场、语旨和语式）功能的贴近。采用贴近翻译（close translation），即使得译文与原文形式、内容、文体更为贴近。按照贴近程度从高到低有七种方法：成分分析、调整译法、描写对等、功能对等、文化对等、同义词替换、释义翻译。纽马克的翻译理论对应用文本的适用性很好，但对文学文本的翻译缺乏一定的解释力。下面结合本章文本和译文加以说明。

语义翻译的应用主要体现在例⑤“A CR network can serve as a robust and efficient communications infrastructure that can solve both the current and future energy management needs of the smart grid. 认知无线电网络可以作为一个强大而高效的通信基础设施，可以解决智能电网当前和未来的能源管理需求。”和例⑥“Unmanned Aerial Vehicle means a power driven aircraft, other than a model aircraft, that performs a useful mission and can be remotely controlled or has autonomous flight capacity. 无人驾驶飞行器是一种除了模型飞机之外，执行有用的任务并且可以远程控制或具有自主飞行能力的动力飞行器。”首先此两个例子中的用词含有诸多专业术语，其中有些表达是人工创新术语，比如“CR”（全称为 cognitive radio），以及“Unmanned Aerial Vehicle”。这些科技英语表达的术语刚开始出现的时候很难在汉语中找到地道的对等表达形式，只能采取近似于顺序直译的方式表达出来，为最高程度的贴近翻译。“CR”译为“认知无线电”，“Unmanned Aerial Vehicle”译为“无人（驾驶）飞行器”，其中前者至今在全国科技名词委员会的术语在线数据库内依然是未定名术语，后者已在 2003 年入选《航空科学技术名词》第一版，并且后者现在有了一个广外人知的俗名“无人机”。此类名词刚刚出现时，最好采用语义翻译的方法，采用与原文贴近程度高的翻译方法，逐字译出，才可保证术语表达的准确性。此两个例子气田部分的翻译也基本采用了与原文贴近程度高的翻译方法，采取顺序直译的方法，把原文内容逐个译出，在个别地方为了符合汉语表达的规范，采用调整译法稍微对句子成分进行了调整。例⑤中的“that”从句在译文中不再作定语成分，而是转换为与原来主句并列的小句。例⑥中的“that”引导的定语从句在英语中的位置是后置的，在译文中调整其位置到所修饰的名词之前。

交际翻译的应用反映在例⑦“But now a team of scientists thinks they

may have found a way to create a compact, graphene-based super capacitor that lasts as long as a battery. 现在一个科研团队已经找到了可以制造出一种以石墨烯为基础,但寿命可以媲美电池的紧凑型超级电容器的方法。"和例⑧"That isn't as good as lithium-ion batteries, which range anywhere between 250 and 750 Watt-hours, but it's still a big jump over current capacitor technology. 它虽然并不像锂离子电池能量密度那么高(锂离子电池能量密度在 250 到 750 瓦时/升之间),但这在电容器科技上仍是一个巨大的飞跃。"交际翻译的目的在于面向读者传递信息,尽量做到简洁、通畅,词汇翻译多采用增译、转译,句子多把被动转换为主动,长句多分译为短句或采取倒译和语序调整的方式译出。目的是让读者会觉得译文的阅读效果如同原文读者阅读原文一样的效果。比如例⑦中对"a team of scientists"没有译为"一组科学家",而是译为更为中文科技读者耳熟能详的"科研团队",两者实属功能对等。同样此句中的"that lasts as long as a battery"没有译为"和电池一样使用时间长度的",而是采取更像同义替换的方式译为"寿命可以媲美电池的"。例⑧中为了译文的明晰化,在"lithium-ion batteries"内容之后增译了"能量密度",而且定语从句的"which"也重译为"锂离子电池能量密度"。这种对原文存在理解困难的内容重新释义解释,使得读者能够更容易理解句子语义。

补充阅读:Earthing

1. What is Earthing?

The earth has been used, almost as long as there have been large power systems, as a reference point from which to measure the voltage of the power system.

The earth, generally, has the same voltage in any geographical area because it is a very good conductor (except for some areas where there are certain non-metallic rock formations).

Equipment connected to the power system is earthed (connected by a metal conductor to ground), to ensure the equipment operates safely and reliably.

2. Why is it Required?

The earthing of power system equipment is important for two main reasons:

- When an insulation fault occurs in a piece of current will be safely diverted into the ground without causing damage to people or other equipment.

- A fault where a live conductor comes in contact with the earth can be detected more easily by protection, allowing the fault to be cleared more rapidly and protecting other equipment in the power system.

3. Earthing

An earth-grid is installed at a substation to limit the rise in voltage of the surrounding ground area during a fault where a large amount is flowing into or out of the earth grid via the earth itself.

The earth grid also evens out the voltage rise around the substation, so that no current flows through people working in the general area of the substation.

4. Transformer and Generator Earthing

The neutrals of power system transformers are grounded to allow out-of-balance earth currents to flow during faults thus limiting voltage rise on healthy phases and permitting selective ground fault relaying to be installed. Supply transformers, generator transformers and distribution transformers usually, but not all, have star connected secondary windings with solidly grounded neutrals and delta connected primaries.

The method of grounding determines how a system behaves during earth faults.

- "Solid" or "effective grounding" results in high earth fault currents and a dramatic voltage drop while the fault is present. The neutral reference remains approximately fixed with respect to the unfaulted phases and therefore there is little rise in the voltage to ground on the un-faulted phases.

- "High impedance system grounding" will result in minimal earth fault current. The phase to earth voltages of the unfaulted phases can rise to the full pre-fault phase to phase voltages as the neutral reference shifts to be equal in potential to the faulted phase potential. In extended systems the system reactance and capacitance can combine to raise the voltage further, sometimes to dangerous levels.

• "Low impedance grounding" is being increasingly used. The phase to earth voltages of the unfaulted phases will rise however they will not rise to the Ml pre-fault phase to phase values during the fault as the grounding impedance is small.

The generators have gone through an evolution of stator grounding techniques. Large generators with their own unit transformers are now almost universally high resistance grounded to limit stator damage by restricting fault currents to a very low level. Smaller generators which are directly connected to 11 kV distribution systems are often low resistance grounded.

The transformer construction vector group is important in determining the effectiveness of system earthing. Transformers with delta/star configuration provide a solid earth point on the star side but none on the delta side. Star/stars transformers with one earthed neutral can provide adequate earthing in some cases; such as for a three-phase transformer where neutral flux can pass through the tank; the tank serves as a delta winding.

Chapter Thirteen Biomass as Alternative Energy

第十三章 生物质作为替代能源

原文

Biomass can be used as fuel, which is as good as other fuel types, e.g. coal, fuel oil and natural gas, etc. It is not difficult to use ①biomass as fuel, but what we need to know is the properties of a given type of biomass and then use an appropriate design of equipment or machinery that suits such a fuel (biomass). Presently, ① biomass has been used as an energy source in household, e.g. as firewood and charcoal for cooking stoves.

1 Biomass as a Substitute for Coal

Coal is the major energy source of cement and paper factories. ⑥ Biomass can be co-fired with coal using the existing manufacturing equipment/ machinery, but there may be a need to modify its structure a little bit. However, whether the factories will be interested to use biomass as fuel or not will depend mainly on the unit price per heating value.

Example 1: If the price of imported coal with 26 MJ/kg of heating value is

译文

生物质可以用作燃料,与其他类型的燃料性能一样好,例如煤炭、天然气等。将生物质作为燃料使用并不困难,但我们需要了解给定生物质的特性,然后采用适当设计的,适合燃料(生物质)的设备或机器进行燃烧。目前,生物质在家庭中已被当作一种能源使用,例如,用于烹饪时在炉具中的木柴和木炭。

1 生物质替代煤炭

煤炭是水泥和造纸厂的主要能源。生物质可以使用现有的工艺设备/机械设备与煤共烧,但可能需要稍微修改设备的结构。但是,工厂是否有兴趣利用生物质作为燃料,主要取决于每单位热值的单价。

例 1:假如进口煤热值为 26 兆焦耳/千克,价格为 2000 泰

2,000 Baht/ton while the price of rice husk is 1,000 Baht/ton. Will the coal price be more expensive than rice husk price?

铢/吨，而稻壳的价格为1000泰铢/吨。用煤发电的成本是否高于稻壳？

2 Biomass as a Substitute for Fuel Oil

2 生物质替代燃料油

Fuel oil is the main fuel of industrial factories which require heat for the manufacturing process, such as the food industry, textile industry, brewing industry and rubber glove industry. Fuel oil is easy to burn, and it is convenient to stock and order. The major machinery used with fuel oil is a fire-tube boiler. The important characteristic is that the heat generated from the combustion process will flow in the tube surrounding with water. The pressure of this boiler type does not exceed 20 kg/cm^2 and the capacity is not over 20 ton steam/ hour.

燃料油是需要热能的工业工厂中制造过程的主要燃料，如食品工业、纺织工业、酿造工业和橡胶手套工业。燃料油易于燃烧，容易储存，订购方便。使用燃料油的主要器械是火管锅炉。其重要的特点是燃烧过程产生的热量会在周围是水的火管内部流动。该类型锅炉的压力不超过20千克/平方厘米，容量不超过20吨蒸汽/小时。

Another boiler type is a water-tuber boiler, which operates at high pressure and produces a significantly greater amount of steam than a fire-tube boiler. The design of a water-tube boiler is opposite to that of a fire-tube boiler, i.e. the heat generated from the combustion process will flow outside the tubes with water circulating inside.

另一种类型的锅炉是水管锅炉，与火管锅炉相比，水管锅炉可以在高压下运行并能产生更多数量的蒸汽量。水管锅炉的设计与火管锅炉的设计相反，即燃烧过程产生的热量将流到管外，而水在管内循环。

However, due to the considerable increase of fuel oil prices since 1997

然而，考虑到燃料油价格的上涨，许多工厂从燃料油转向生

factories switched from fuel oil to biomass because the economic crisis made biomass cheaper in terms of "Baht/heating" value.

Example 2: ③If the heating value of fuel oil is 41.1 MJ/liter and the combustion efficiency of the fuel oil boiler is 90%, determine the required amount of rubber wood tops (in kilogram) to be used as a substitute for 1 liter of fuel oil, given the combustion efficiency of the rubber wood chips boiler at 70%.

Example 3: From the above example, ⑦if a factory that used to consume fuel oil at 50,000 liter/ month at the price of 15 Baht/ liter and has switched to use rubber wood tops, of which the price is 800 Baht/ ton, determine the fuel cost saving gained by the factory from the fuel switching.

物质燃料,这是因为1997年的经济危机使得以"泰铢/热能"计算的生物质价格更便宜。

例2:如果燃料油的热值为41.1兆焦耳/升,燃油锅炉的燃烧效率为90%,在已知橡胶木屑锅炉的燃烧效率为70%时,可以确定替代1升燃料油所需的橡胶木木条用量(以千克为单位)。

例3:从上面的例子可以看出,如果一家工厂以前曾每月要消费50 000升价格为15泰铢/升的燃料油,并且已经改用价格为800泰铢/吨的橡胶木木条,可以确定该工厂的燃料转换后节省的燃料成本。

3 Biomass Gasification for Electricity Generation

Theoretically, in order to supply electricity to the grid, the construction of a large power plant would be preferred because a large power plant requires smaller construction cost per MW and cheaper maintenance cost when compared with a small one. However, in the case of a biomass power plant, the plant capacity is usually not more than 10 MW due to the limitation of fuel supply. ⑤A large plant capacity will need a large amount of fuel,

3 生物质燃料煤气化发电

理论上,为了向电网供应电力,大电厂是更受欢迎的,因为对比小电厂,大电厂每兆瓦的建设成本更小、并且保养的成本也更低。然而,对于生物质电厂,由于能源供应的限制,发电厂装机容量一般不超过10兆瓦。大容量发电厂将需要很多的燃料,这就增加了生物质燃料供应和进一步燃料来源收集方面的正本,并且因此也增加了更多的运

which will increase the costs of fuel biomass supply and collection from farther sources, and hence higher transportation costs too. In contrast, for a natural gas power plant, natural gas is transported via pipelines. For a coal-fired power plant, a large amount of imported coal will be shipped, or the plant can be located in the vicinity of a coal resource. For this reason, fossil-fueled power plants are usually much ⑤ larger than biomass power plants.

输成本。相反,天然气发电厂由管道运输。火电厂需要的大量的煤炭可以运来,或者可以将这种火电厂建在临近煤炭资源的地方。出于这种原因,化石燃料电厂比生物质燃料电厂的容量更大。

Gasification is a power generation technology for small capacity (not exceeding 1 MW). The process starts with biomass digestion so that the size of biomass fuel is not more than 10cm. Biomass is then delivered to the combustion chamber where the amount of air inflow will be limited, which will cause incomplete combustion. The result of the process is mainly carbon dioxide with average heating value of about 5 MJ/m^3; this can be more, or less, depending on the technology used.

气化法是不超过1兆瓦的小容量电厂的一种发电技术。这一过程从不超过10厘米的生物质燃料的细菌分解开始。然后,燃料将传送到不完全通风的燃烧室进行不完全的燃烧。这一过程中产生的主要是二氧化碳,平均热值在每立方米5兆焦。这一数值的多少取决于使用的技术。

The derived gas mixture is called "synthesis gas" which can be used as fuel for ②direct heating, such as ②rice paddy drying. If the synthesis gas is used for electricity generation by a diesel engine, tar must first be eliminated. Using the synthesis gas can reduce diesel oil

导出的混合气体叫作合成气,可以作为燃料直接供热,比如用于水稻干燥。如果合成气被用来作为柴油机发电的燃料,一定要首先剔除焦油。使用合成气能够减少75%柴油的消耗或者完全替代柴油,但是发电量

consumption by 75% or it can completely replace diesel oil, but the generation capacity will considerably decrease. The electricity generation efficiency of the gasification system ranges between 20%～30%, depending on the technology, design and efficiency of the equipment used.

将大幅减少。根据技术、设计以及设备的有效使用，合成气系统发电的效率在20%～30%之间。

④ Biomass to be used as fuel for a gasifier should have an appropriate and consistent size with no more than 20% moisture, such as rice husks and ground wood chips, palm shells and corncobs. Sawdust is not suitable for a gasifier because the size of sawdust is too small and hence air cannot flow through it, while biomass of which the size is too big, such as not-yet-ground wood slabs, is not suitable for gasifier either as some will not be burned.

被用来作为气化炉燃料的生物质，就像燃烧秸秆、碎木片、棕榈壳和玉米芯，应该具有恒定和适当的大小以及不超过20%的水分。锯末不适合气化炉的原因是锯末太小而不易通风。太大的生物质也不行，就像未研磨好的木板也不适合气化炉，因为不能充分燃烧。

4　Constraints of Using Biomass as Fuel

4　生物质燃料的限制

Biomass fuel is different from fossil fuel in various aspects, i.e. biomass fuel is in a variety of forms; it has scattered resources; and its quantity in each year depends on the climate. Therefore, in utilizing biomass as fuel, the following should be taken into consideration:

生物质燃料与化石燃料在很多方面存在差异，例如，生物质燃料形式多样化；生物质燃料资源具有分散性；生物质燃料的年产量取决于当年的气候条件。因此，利用生物质作为燃料时，应考虑以下因素。

(1) Biomass is light in weight, so it requires several trips of transportation if loaded by truck. In addition to the

(1)生物质质量轻，体积大。如果用卡车装载，需要多次运输，除了会增加运输成本外，还

increased transportation cost, it may affect a community in terms of traffic jams and road safety problems.

会影响社区的交通和道路安全。

(2) Most biomass contains high moisture. The combustion chamber for biomass has to be large. If the moisture content is very high, the installation of a dehumidified device may be required, resulting in an increase of investment cost.

(2)大多数生物质含有高水分,因此生物质的燃烧室必须很大,如果其水分含量非常高,则可能需要安装除湿装置,从而导致投资成本的增加。

(3) The ash of some biomass has a low melting point. Therefore, too high temperature in the combustion chamber will cause the ash of biomass with low melting point stick to the surface of combustion chamber and hence reduces efficiency of the steam boiler.

(3)某些生物质的灰分熔点较低,因此燃烧室温度过高会导致低熔点生物质的灰分黏附在燃烧室表面,从而降低蒸汽锅炉的效率。

《术语表》

biomass	生物质
heating value	热值
fuel oil	燃料油
gasification	气化
fire-tube boiler	火管锅炉
synthesis gas	合成气
water-tuber boiler	水管锅炉
MJ	兆焦耳(megajoule)
dehumidify	除湿

翻译难点讲评

① 专业词汇中包括有常用词汇的专业化,或者也称之为半科技词汇,使用频率较高;还有一种是叫作纯科技词汇,这些词的意义狭窄,通常只表示一

种专门的科学或技术定义。此处段落内出现的3处“biomass”有两种可能的译文“生物质,生物量,生物质能”。《电力名词》第二版:一切直接或间接利用绿色植物光合作用形成的有机物质。包括除化石燃料外的植物、动物和微生物及其排泄与代谢物等。《植物学名词》第二版:在一定时间内,生态系统中某些特定组分在单位面积上所产生物质的总量。《微生物学名词》第二版:泛指某一系统中的一切生物物质的总量,即单位面积或单位体积中生物物质的质量数。根据词义选择的专业相关原则,这里选择“生物质”最符合要求。

② 科技翻译原文译文功能对等在词汇层面实现意义对等的翻译方法有词类转化。比如这句中包含有的两个动名词的形式,“direct heating”和“rice paddy drying”。词类转换法就是在翻译中对原文的词性及句子的各种成分进行转换,比如名词转化为动词。如这句话如果直接按名词来译的话,意思可理解,但不符合汉语习惯。根据英汉语之间的差异,英语是静态语言,语言中更喜欢用名词,而汉语中则为动态语言,多用动词,所以将这话句中的动名词的词性进行转化,变为汉语中的动词。这样既符合中文的习惯,读起来也更通顺了,译为“The derived gas mixture is called ‘synthesis gas’ which can be used as fuel for direct heating, such as rice paddy drying. 导出的混合气体叫作合成气,可以作为燃料直接供热,比如用于水稻干燥”。

③ 科技英语客观、简练、逻辑性和科学性强等特点,非谓语动词是其主要的实现手段之一。另外,长难句太多容易造成句子冗长,并且谓语动词过多,因此多借用非谓语动词。这些非谓语形式,在句中可做主语、宾语、表语、定语、状语。比如这句“If the heating value of fuel oil is 41.1 MJ/liter and the combustion efficiency of the fuel oil boiler is 90%, determine the required amount of rubber wood tops (in kilogram) to be used as a substitute for 1 liter of fuel oil, given the combustion efficiency of the rubber wood chips boiler at 70%.”就包含了多处非谓语动词的使用,如“...to be used...”和“..., given...”,分别做后置定语和状语。翻译的时候,当英语句子的叙述逻辑关系与汉语基本一致时,可以按照英语原文的顺序翻译成汉语,即可以采用顺译的方法。如果英语和汉语在语法结构上有着许多差异,则可以根据意义灵活地采用省译、增译、语序调整、词性转译、句子成分转译或者分译的方法处理。本句中的“..., given...”实际表达

的是如果此处条件已知，可确定相应的其他值。可以采用调整语序，把此状语放在主句的前面，以符合汉语的逻辑表达规律。此处“... to be used...”和前面的“... the required...”所表达的含义相同，可以省译，其他的后置定语内容翻译为前置定语。最后译为“如果燃料油的热值为41.1兆焦耳/升，燃油锅炉的燃烧效率为90%，在已知橡胶木屑锅炉的燃烧效率为70%时，可以确定替代1升燃料油所需的橡胶木木条用量(以千克为单位)。”

④ 此段的两个句子字数都超过了30字，属于长句，其中第二句更加复杂。第一句的主语和宾语的后面都有较长的后置定语，第二句使用了较多的从属连词，包含有定语从句和状语从句等几个类型的从句。对于长难句的翻译，主要有以下几种翻译方法：增译法、省译法、分译、合译、顺序法，倒译法和结构调整。第一句“Biomass to be used as fuel for a gasifier ... with no more than 20% moisture, such as ....”里面的后置定语按照汉语的表达习惯一个调整为前置定语，另外一个定语成分调整为宾语(此处变成了与先行词并列的宾语)，举例成分的位置调整到汉语谓语动词之前。此句译为“被用来作为气化炉燃料的生物质，就像燃烧秸秆、碎木片、棕榈壳和玉米芯，应该具有恒定和适当的大小以及不超过20%的水分。”第二句“Sawdust is not ... because ... and hence ..., while biomass of which the size is too big, such as not-yet-ground wood slabs, ... either as some will not be burned.”在转折连词的地方进行拆分，译为两个单独的句子，分译后的句子都采用顺译的方法，按照原句中的语序和逻辑关系译出。此句译为“锯末不适合气化炉的原因是锯末太小不通风。太大的生物质也不行，就像未研磨好的木板也不适合气化炉，因为有些不能燃烧。”

⑤ 科技英语中使用的大量词汇是普通英语词汇，这些词汇都是多义词，需要从上下文的关系中来探求正确的译法。而且客观性科技文献翻译要做到不使人费解，更要做到不让人误解。此句“For this reason, fossil-fueled power plants are usually much larger than biomass power plants.”中的单词“large”如果简单地理解为“大”，那就是没有获得其上下文中的具体含义。此段上句有处也出现了该词，“A large plant capacity will need a large amount of fuel, ...”。由此可看出这里该词应该是指电厂的装机容量大。该词“large”的词义需要仔细推敲，进行修辞性增译，结合上下文语义、语境，译文必须增添新词才能准确达意，这既是原文内涵的要求，也是译文

行文的需要。两句分别译为:"出于这种原因,化石燃料电厂比生物质燃料电厂的容量更大"和"大容量发电厂将需要很多的燃料……"

翻译理论及应用讲评:语境理论

语境影响科技翻译中理解的准确性和表达的通畅程度。自从1884年德国哲学家在《算术基础》中提出语境(context)原则以来,语境研究逐渐受到重视。语境对话语理解上的重要性首次提出者为波兰人类学家马林诺夫斯基(Malinowski)1923年写给奥格登(Ugden)和理查兹(Richards)所著的《意义的意义》(*The meaning of meaning*)一书的补录中。马氏主要关注情景语境(situational context)和文化语境(cultural context)对于话语理解的作用。英国系统功能学派的创始人弗斯(J.R. Firth)把马氏的语境归入非言语语境,同时又扩充增加了语言语境的(linguistic context)的概念,初步形成了比较完整的语境理论。弗斯的学生,伦敦功能学派的代表人物韩礼德1964年把语域(register)引入语境理论。韩礼德1978年的著作《语言作为社会语义符号》(*Languages as Social Semiotic*)给语域的理论定义表明其就是用语域代替了语境,分为语场(field)、语旨(tenor)、语式(mode),实质上强调的是语言使用,属于非言语语境。美国社会语言学家海姆斯(D.H. Hymes)在1967年和1974年也提出了语境相关理论,他以SPEAKING字母缩略词代表了语境的基本构成八个因素:情景(situation)、参与者(participation)、结局(ends)、行为特点(act sequence)、基调(key/manner)、媒介(instrumentalities)、交际规则(norms of interaction)、风格(genres)。该观点与韩礼德类似,都把语境的构成要素与语言的三个元功能对应,比如语场对应概念功能,语旨对应人际功能,语式对应语篇功能。以上学者们的研究加大了语境构成要素研究的深度(语言语境和非语言语境两个层次),又扩展了语境功能特征研究的广度(多个功能)。Georgakopoulou(1997:18)进行了简明全面的概括,明确讲语境分为四种类型:情景语境、文化语境、语言语境和认知语境。

由此,语境对词义的制约作用是显而易见的了,词或语义成分都不孤立而存在,其含义都要通过语境来确定。科技英语翻译中的语义转换必须首先基于语境分析,先从下而上地分析具体的上下文语言语境,比如形态语义和语法和句法特征,再自上而下分析影响科技翻译的文化语境,比如社会、历史、哲学、物质文化、宗教文化、语言文化、精神文化、政治文化和地理文化,同时也要考虑译者本身所处的客观经验世界，如译者的专业知识、百科知识和工具技能等各种影响翻译的认知能力，以及译者所处的交际现场环境等。以下举例说其应用。

例⑥原文“Biomass can be co-fired with coal using the existing manufacturing equipment/ machinery, but there may be a need to modify its structure a little bit.”。译文为“生物质可以使用现有的工艺设备/机械设备与煤共烧，但可能需要稍微修改设备的结构。”此例英语原文提供了丰富的语言语境。例如“co-fired”单词的两个表达具体意义的语素“co”（一起，协同，共同）和“fired”（燃烧）。虽然“co-fired”并不是一个由全国科学技术委员会定名的术语，但是根据其形态结构提供的语境，依然可以准确把握其词义。最后译文没有使用“一起燃烧，共同燃烧”，而是采用了表义相同，但形式更简洁的“共烧”的译文。虽然牺牲了一点其理据性，但依然对等。该句中的“its”是个替代词，需要在语境中仔细判断其所替代的内容。本句的句法语境含有内容“existing...”，此处就是暗含意思为“未修改其结构时，使用现有的设备”。另外，从更大的上下文的语篇语境中，也有提到“Biomass to be used as fuel...consistent size with no more than 20% moisture...”，可以推论“生物质燃料只需在大小和水分含量上满足要求即可”。而这些方面并非改变生物质的结构。因此此处“its”替代的应该是“manufacturing equipment/machinery”这个表达形式。进一步分析这个表达形式里面的搭配关系，“manufacturing”所修饰的词到底是后面的一个还是两个词，以及该词的含义是否是“生产，制造”。查找全国科学技术委员会的术语在线网站，发现“manufacturing equipment”是2003年在《机械工程名词第二分册》第一版的定名术语，译为“工艺设备”，指的是“完成工艺过程的主要生产装置”。该术语在线没有收录“manufacturing machinery”这个术语。据此，该表达形式译为“工艺设备/机械设备”，而代词“its”依据此指代内容也可简化为“设备的”。

例⑦原文“...if a factory that used to consume fuel oil at 50,000 liter/month at the price of 15 Baht/ liter and has switched to use rubber wood tops, of which the price is 800 Baht/ton, determine the fuel cost saving gained by the factory from the fuel switching.”。译文为“……如果一家工厂以前曾每月要消费50000升价格为15泰铢/升的燃料油，并且已经改用价格为800泰铢/吨的橡胶木木条，可以确定该工厂的燃料转换后节省的燃料成本。”该句中的翻译难点是“rubber wood tops”。在词典里并无该表达的译

文。该句子内容含有专有的单位名称"Baht",即采用泰铢计量的货币单位名称。此处提供了非常有价值的文化语境信息,即该木材为泰国的知名或常见木材类型。在网络上搜索知道了泰国以出产橡胶木和酸枝木著称。"Rubber wood"必定指的是其中之一的"橡胶木"。那么"tops"又指什么?根据此句所在文本的语域(对应指的是语境),其语式(对应指的是语篇功能)为介绍燃料油切换为生物质燃料后降低了燃料成本。即首先,"rubber wood tops"是属于生物质燃料这一语义范畴的。根据整个语篇提供的信息,生物质要汽化后才可使用,而且被用作气化炉燃料的生物质不能太大,比如木板就太大不易放入炉内燃烧,也不能太小,比如锯末就太小,不通风不能燃烧。所以此处的"tops"是大小适当的木料类型,可译作"木条"。

综上所述,依据语境理论进行科技英语翻译是一种言语交际的过程。这里不仅是语言解码,还要求译者密切联系各类语境,使用相关语境信息进行推理和判断,才能保证译文的准确的通畅,避免翻译错误和翻译腔。

补充阅读:Lightning and Surge Protection

1. Lightning Protection

As most power systems have a network of overhead wires and tall metal structures carrying those wires, the system is prone to lightning strikes. A lightning strike to any conductors in an electrical circuit causes a surge in the power flow, and can lead to cascading failures of equipment along the circuit.

Lightning protection is used to guard against this by diverting the lightning strike directly to ground, without flowing through the main circuit.

A lightning protection system comprises three parts:

• An air termination system, which consists of wires placed above the power conductors to capture the lightning strike;

• Down conductors, which are the steel towers and/or conductors that take the current from the overhead wires to ground;

• Earthing system, which allows the current to flow from the down conductor into the ground. The earthing system can be an earth-grid, the reinforcing bar used in concrete, earth pin, etc.

2. Surge Protection

Surge protection is commonly installed near transformers to prevent a power surge in the transformer, which can involve lengthy repair times and significant expense. Surge protection is commonly installed near transformers to prevent a power surge from causing an insulation fault in the transformer. They are also installed on the lines at the entry to substations to prevent damage to instrument transformers. Surge arrestors are a form of 'pressure release, for a power system. They allow current to flow to earth through them if the voltage becomes too high. Surge arrestors are very used in all kinds of damage.

Surge arrestors may be installed on a power system to protect equipment against voltage and current spikes that may flow along a circuit due to a lightning strike (should it not be adequately protected by lightning protection), or switching surges.

Chapter Fourteen Reference Translations for Supplementary Readings

第十四章 补充阅读参考译文

第一章补充阅读参考译文:汽轮机

有关汽轮机和辅助系统的其他要求,请参阅附件 C 中“系统定义”中的相关文件。

蒸汽涡轮机应为适合室内安装的单轴串联复合式 3000 RPM 单一再热冷凝式。承包商应选定发电站的蒸汽循环布置,包括涡轮发电机参数,以便在额定主蒸汽流量压力和温度条件下,在发电机升压(generator step-up, GSU)的高压端实现额定净电输出。

蒸汽涡轮机在汽阀全开(valves wide open,VWO)的情况下,应在此点以上有 3%(最小值)的流量余量。涡轮发电机和其他电厂设备应能在汽阀全开和常压条件下连续长期运行。循环参数包括给水加热器数量、进口端和末端温差、加热器排水布置、背压、末级叶片选择和其他蒸汽循环参数应由承包商根据资本成本和运营成本相对值度量比进行优化。

汽轮机应是由高压汽轮机(HP 汽轮机)和中压汽轮机(IP 汽轮机)壳体和四或六个流低压汽轮机(LP 汽轮机)壳体向下排入表面冷凝器组成的串联复合再热机。

涡轮机壳体应沿着涡轮机转子中心线水平分开。

涡轮机转子应该静态和动态平衡。应提供完整的阀门保护系统以保护涡轮机过速。

减少和控制涡轮机热应力的设计特征应包括:

1. 启动时全弧进气能力仅使用控制阀配置控制完整的弧线准入能力，不使用涡轮截止阀。

2. 滑动油门压力操作按钮将压力从最小负载调整到100%MCR（maximum continuous ratings，最大持续功率）状态。然后将压力固定在VWO（103%MCR）状态条件下。

3. 转子应力监测系统通过监测和显示在蒸汽与金属温度匹配、加速和负载期间的循环寿命消耗提供监控和根据转子应力来协助启动。

涡轮叶片和涡轮发电机控制系统应设计成允许与附件A中列出的电网规范要求相符的低频和高频运行。涡轮的启动和初始加载应在原始设备制造商(OEM)的推荐程序操作指导下，由涡轮机控制和仪表控制系统通过自动或全部手动方式进行控制。

所有的跳闸保护功能都应用冗余传感器来实现，以避免因一个设备发生故障而导致涡轮系统无法运行。

应使用自诊断系统来识别超出正常运行范围的设备。自诊断系统应该对超出运行范围的设备发出警报，以便对其识别并维修。

涡轮机制造商应确定涡轮发电机针对一些条件的功能约束，包括启动和装载过程，速度，冷凝器压力，节流和再热温度和压力，蒸汽质量等工况。涡轮机制造商也应提供在冷、半热和热条件下涡轮机的启动和运行曲线。

涡轮机机油系统至少应包括：一台轴驱动的主油泵和一个交流油泵，一个直流应急油泵，两个全容量双联机油冷却器，一个安装在油箱上的蒸汽抽油机，一个油箱（净机油油箱），机油调节器和一个脏机油油箱。

主机油油泵应该由涡轮轴驱动，并配有一个备用的全流量交流电机驱动泵。还应配置系统运行所需的所有其他泵。所有电机驱动的机油泵应安装在机油油箱上。每个油泵的吸油和回油管路应设有过滤器。

涡轮机的紧急油泵上不应设置紧急停止按钮。

所有的机油管道应使用18-8不锈钢（18%铬加8%镍），并采取保护，保护管可以使用碳钢。设置机油管道时，应避免穿过或位于任何蒸汽管道的正上方。机油管道应倾斜以避免气穴，并尽可能使用重力引流，同时机油管道应尽量在涡轮机底座的范围内。应设置冗余压力变送器，以便自动顺序启动电动油泵。应设置单独的泵排放压力变送器指示每个油泵的运行。应规定要从控制室测试交流轴承油泵和直流应急油泵自动启动。还应为公共机油供应头和每个轴承的单独机油供应配置压力变送器。

应设置涡轮机油净化系统，它包括循环泵、压力颗粒过滤器和聚结除水过滤器。该系统应去除10微米及以下的颗粒物质，并去除机油中的所有游离水。该系统应设计为每小时至少处理20%的系统容量机油。涡轮机机油系统还应配备一个净机油油箱和一个脏机油油箱，每个油箱的工作容量等于涡轮机机油油箱的总容量，并配有一个液位变送器用于向控制室指示液面。

涡轮排气罩喷水系统应该是一个完整的系统，应达到供应商的标准，防止其温度过高以保护排气罩。

轴封系统和汽封冷凝器应达到涡轮供应商的标准。辅助蒸汽系统提供密封蒸汽，在密封蒸汽的给水阀和卸荷阀周围应设置旁通阀。

第二章补充阅读参考译文：锅炉

系统设计必须满足工厂的所有要求，锅炉应基于超临界技术，应为先进设计的滑压运行直流型锅炉，采用单再热、单炉膛、两个火球，配备省煤器和再生空气加热器、垂直管布置炉壁、均匀通风，采取悬架式钢结构框架结构。

锅炉应是室外、粉煤燃烧、均匀通风、直流式、配备有省煤器和再生式空气加热器的机组。锅炉的设计、建造和性能测试应符合ASME锅炉和压力容器规范、ASME B31.1、NFPA和工厂和机械法1967年（第139号法令）和法规的要求。

为了控制结渣和积垢的影响，承包商应为锅炉的各个部分提供一个完整的吹灰系统，并配备详细图纸显示出设备各部位、吹扫角和炉壁上的任何备用孔。

锅炉结构应该装备一个不保温的锅顶和滑动壁板，从锅炉顶部测量向下达10米。锅炉应是一个1×100%容量的装置，其设计单独使用或组合使用业主要求的煤炭规范中所列的煤，采用制造商的标准低NO_x燃烧器和燃尽风装置。燃烧空气应由主送风机和强制通风机提供。锅炉应具有顶部支承，并应由承包商确定其双回程设计。

锅炉和辅助设备的设计应能够从启动到锅炉最高连续额定功率（BMCR）时安全可靠。在不使用补充燃料的情况下，从30%的负荷到BMCR该装置应能够稳定地燃烧。启动燃料为馏出油。电站的设计应确保可在受控、稳定的模式下连续安全运行，以便在无馏出燃料辅助的情况下从最低负荷运行。锅炉的最大持续功率应相当于MCR的105%。

锅炉主蒸汽温度应根据制造商的标准，由燃烧器倾斜和减温喷淋水的组合控制。如果主蒸汽的温度控制是由降温喷雾提供的，那么减温器应位于一级和二级过热器之间，以及二级和三级过热器之间。

再热蒸汽温度应通过倾斜燃烧器控制。或者用第二种方法采取减温喷淋水来实现高温再热蒸汽温度控制。

末级高温过热器或高温再热器不允许使用TP347H。末级过热段适用TP347HFG、304H、Super 304H、NF709和HR3C，并进行内喷丸处理，以提高抗剥离性。过热器和再热器设计应提供足够的流量混合，以确保所有负载下的温度均匀，并且集热器、管和支撑件应具备热膨胀性。所有各段组件应可清洗、检查、维护和拆卸。

炉膛水墙应采用垂直(膛线)管的隔膜结构。

锅炉的设计应能在满负荷范围内令人满意地运行，煤的性质在表2-2版本C中，在雇主要求的第2.3节(燃料)和附件B3燃料煤的列表中。锅炉制造商应设计锅炉及相关辅助设备，使其在满负荷范围内可靠运行能使用高水分煤(按重量计高达30%的水分)。在选择磨煤机设计时，应考虑煤的含水率。此外，锅炉相关的风机(包括密封风机)应对额外的水分进行处理。空气加热器应考虑到需向磨煤机提供一次热风用以干燥比较湿的煤。

炉膛的尺寸应确保炉膛内的燃料完全燃烧，而不产生对水冷壁、过热器或再热器元件的火焰冲击。炉膛内不得使用分隔壁。所有的辐射水冷壁表面都应安装在炉子的壁、顶或底上，而不需要把水冷面凸进炉内。凸进炉膛内的蒸汽冷却板应宽间距设计，便于运行期间的清洗。充足的空间将为炉的所有侧面提供吹灰器覆盖。应提供足够的空间来维护吹灰器。

锅炉每一个计划区域的热量输入，燃烧器区域热释放(切向或前和/或后烧设计)，每平方米有效投射面的炉子可用热量，进入突进隔板的最大热气体温度和最大热气体速度，都应该由承包商来选择，以确保锅炉在锅炉最大持续额定功率(BMCR，常压条件)下，使用推荐范围内的燃料，结渣、结垢和腐蚀方面的稳定和可靠运行。对于非本地煤，飞灰中的碳不超过5%，对于含灰分大于2%的本地煤，不超过3%。在整个负载范围内，对于任何燃料而言，炉膛出口气体温度应小于还原环境中的最小初始变形灰熔融温度。

承包商应在锅炉尾部烟道的详细设计过程中提供烟气流速计算，选择和设计适合于预期用途的材料，以确保高热气体流速已被永久固定结构消除。

承包商应采取适当的措施，如气路分隔，防止管束振荡和噪音。整个锅炉的气体速度不应超过每秒 18 米。

承包商应提供管泄漏声学检测系统，其中包括检测器、电子设备和控制室指示器。

在电厂和设备维修所需的其他特种工具中，承包商应提供一个用于锅炉维护的动力操作维修支架。

维修支架的设计应使部件能够通过炉膛检修孔/人孔，并能在炉内完整装配使用。

高温区的炉膛材料适合于预期功能与服务。

应进行循环分析，以确保水墙管、集水管、降液管和进料/释放管的正确设计，其分析结果应与业主单位分享。

过热器和再热器管应该是无缝连接的。省煤器应该是非蒸型，省煤器管应该是无缝以及裸露在外的，无散热片。

在详细设计中，承包商应考虑炉内材料的选择应适合于预期功能与服务，同时还需要考虑到热流和可能的流动加速腐蚀作用。

锅炉系统（包括炉膛，以及从空气入口到烟囱的所有烟道和通风管道）应按设计有符合 NFPA 85（National Fluid Power Association，美国全国流体动力协会）要求的正、负压力。

机组的计划运行是满负荷情况下的基础负荷运行。但是，锅炉压力部件和其他相关设备的疲劳分析应以循环机组的运行为基础（每天满负荷运行 10～16 小时，每天其余时间相对应于最小负荷不使用补充燃料降负荷运行）。

还应提供锅炉充水系统。

直流式锅炉在启动和低负荷运行时应具有分离水和蒸汽的功能；应设计使容器和内部部件在从点火到完成一次操作的所有负载上都具有足够的分离能力，而不需向过热器喷水。

将提供锅炉水再循环泵。在正常和所有可能的瞬态条件下，考虑到压力和温度的一般条件，泵的尺寸应能足够处理启动和循环过程中的最大流量。

泵的安装应有益于维修、修理、拆卸和调整的通道。

承包商应在省煤器和空气加热器之间的锅炉尾部烟道气体空间中，均等安装至少 8 个氧探头。

承包商应提供锅炉和相关设备的短、长期封存系统。承包商应确保湿法储存系统与他安装的计量设备兼容。

应提供红外测温探头，以测量炉膛出口处的烟气温度。

应提供一套完整的吹灰系统，以便从整个锅炉装置的传热表面有效清除各种形式的灰粉。吹灰器应为电动机驱动装置。应提供手动旋转和收回吹灰器的方法。所有元件应由一端封闭的无缝钢管构成。用于去除灰渣沉积物的吹灰系统应配备可编程控制器，包括单个或多组鼓风机的可变组选择以及用于炉管温度监控的热通量传感器。吹灰蒸汽集箱应配备自动排水装置以去除冷凝水。用于对流和过热表面和炉壁的吹灰介质应为蒸汽。每个吹灰器应配备隔离阀，以能够在系统运行时进行维护。

空气加热器和锅炉辅助设备的类型应取决于设备供应商的选择和相关的原始设备制造商标准。空气预热器设计和材料选择应适合于减轻酸露点腐蚀问题。

炉膛的设计压力应为 ± 533 毫米汞柱(mmhg)。

所有燃烧空气管道应设计为 - 203 毫米汞柱(mmhg)，以及相当于运行期间任何位置(包括主燃料跳闸)可产生的最大瞬时压力加上余量的正压。

所有烟道管道至引风机入口的设计应符合正压 203 毫米水柱，和相当于在运行期间任何位置(包括主燃料跳闸)的最大瞬时负压加上余量的负压。

设计压力不得低于 NFPA 85 要求的压力。

锅炉管理控制系统应安装在符合 NFPA 85 要求的 DCS 处理器中。锅炉应提供燃烧器管理控制系统所需的仪器仪表，包括主蒸汽压力和温度变送器，再热压力和温度变送器，燃烧气流和温度变送器以及氧化锆电池式燃烧气体氧分析仪。最终主蒸汽温度控制应使用减温喷水控制阀来完成。通过气动控制阀调节吹灰器系统的蒸汽供应。应配备一组锅炉管金属温度热电偶，并连接到数据传输系统(Data Communication System, DCS)以获取操作信息。

炉膛下部两侧应设置宽 813 毫米、高 1397 毫米的水冷检修门。过热器、再热器、拱管与套管之间，省煤器管束之间以及任何其他需要进入的区域应配备宽 457 毫米、高 610 毫米的耐火材料内衬检修门，以允许检修。所有的门都应该有铰链和压力密封。应配备观察孔，以便清楚地看到燃烧器附近的每个燃烧器的侧壁，前壁和后壁以及灰斗斜坡。在运行过程中需要检查的任何其他区域应同样安装。这些孔应位于离临近楼层或平台层不超过 1.5 米的位置，无须搭建脚手架，方便进入。

第三章补充阅读参考译文:发电机

发电机的输入轴由一个“原动机”(如柴油发动机,或蒸汽机、风机或水轮机)驱动。原动机为发电机的输入轴提供机械能。在发电机内部,输入轴会转动一个“转子”,产生旋转磁场。旋转磁场会穿过发电机外壳中的“定子”(固定或静态)绕组(线圈)。这些绕组会连接到发电机的输出。变化的磁场通过“磁感应”过程在发电机输出中产生电压。

当发电机输出(通过传送系统)与一个负载连接时,电流则将电能传送给负载。

电力系统中被普遍使用的发电机有两种主要类型:同步发电机和感应发电机。绝大多数发电机都是同步发电机。电动机有各种对应类型;然而同步电动机十分少见,并仅被使用在特殊应用设备中。

1.同步发电机

如其名称所示,一台同步发电机要求转子以相对于电力系统频率的固定速度旋转。转子旋转频率可以与电力系统相同(比如定额转速3000转/分)或者降低至另外的旋转频率,此时额外的转子线圈被称为“极(凸极或隐极)”,并且原动机可以以其他转速转动。

转子有一个由稳定(直流)供电的绕组。恒定电流形成与转子一起旋转的磁场。一台同步发电机可以通过改变流入转子绕组的电流量来控制输出电压。从历史上看,这通常是在电力系统上提供快速电压控制的最廉价的方式,所以几乎所有发电机(和所有大型发电机)都已经是同步型发电机了。

2.感应电机

感应发电机不是锁定在电力系统频率上,虽然在正常情况下,它们运行在一个小的速度范围内。它们的主要优点是其机械简单,制造成本低。然而,它们对电力系统性能有不利影响(与同步电机或其他类型的载荷相比)。两种常见的感应电机是“鼠笼式电机”和“双馈感应发电机(DFIG)”。

鼠笼式感应发电机是因为其采用圆筒形转子笼而命名的。在转子内部,类似于松鼠运动的笼子或鼠标轮。

转子不需要电源来维持磁场。相反,通电的定子绕组感应电流沿着旋转的转子杆流动,形成磁场。转子对于速度的变化非常宽容,因为转子上存在的磁场不是被保持在固定位置，而同步发电机中由外部电源维持的磁场位置固定。

双馈感应发电机(DFIG)像普通感应电机(如鼠笼式)和同步发电机的组合运行。转子上的三相馈电逆变器允许转子的磁动势(MMF)独立于机械旋转而旋转。当转子速度变化时，用合适的逆变器控制系统，转子磁动势与电力系统频率保持同步。双馈感应发电机在风力发电中非常有用，因为叶片可以在一定的转速范围内旋转。风力涡轮机的空气动力效率在一个特定的转速下是最大的，并且最佳转速随风速变化而变化。因此变速运行最大限度地提高了风能捕获效率。此为，与鼠笼感应电机不同，双馈感应发电机能控制系统电压。但是，相对于发电机尺寸，逆变器太小，因此电压控制的范围和数量都是有限的。

3.逆变器连接

逆变器是一种电子设备，可用于将发电机连接到电网。逆变器从直流(DC)输入或从不同频率振荡的输入，输出到电力系统一个交流(AC)。这允许输出直流(例如光伏系统)的电源连接到电力系统(使用的是交流)。

逆变器还可以使同步电机以可变速度运行。与双馈感应发电机相比，可以使用机械简单的装置(避免使用高功率滑环)，只要逆变器的额定功率值没有比主机输出功率小。

即使当能源是不可用时，逆变器可以提供极好的电压控制。

4.讨论

在实践中，电力系统上的大多数发电机是同步的(并且可调节电压)，并且大多数负载是感应电动机(其余的是电阻性的或静态的)。理论上，这一点有可能被逆转，即存在感应发电机和同步电动机负载的电力系统(具有负载调节电压)。甚至这一存在被认为是更好的，因为负载电压是最受关注的，并且最好是在负载附近进行调节。

然而，发电机很少，并且它们有很多负载。一般来说，在少量发电机上使用价格昂贵的(并且有点复杂的)电压控制设备是更省钱和更有效的。原则上，不必在每个发电机上提供这种控制。在技术文献中已经提出使用大型感应发电机的建议，然而，这在实践中并没有实现。

许多早期风力发电机使用感应电机。因为，它们更便宜，更重要的是，在涡轮机舱中需要重量更轻、机械更简单、维护更少的机器。当机器和风力发电场相对较小时，这是可以接受的，但他们现在在接近或超过传统电厂的规模。

鉴于上述情况，感应发电机正越来越多地被要求安装辅助电压控制设备，以克服发电机的固有缺陷。

第四章补充阅读参考译文：发电与能源

通过电力系统提供的电力质量不仅取决于是否有足够的发电量、适当的备用余量，还取决于用于发电的能源类型。在考虑发电机对系统稳定性的影响时，对能源的物理控制进行建模是很重要的。在某些情况下，可以定制一个发电单元的设计，以适应系统所需的稳定特性。

1. 热力涡轮机

用于热发电的能源可以是燃料（如煤、天然气、生物质或任何可以燃烧产生热能的材料）或其他热源（如核能、地热或太阳能热能）。所产生的能量被蒸汽涡轮机、燃气涡轮机或组合涡轮机捕获并转化。

对于蒸汽轮机，能源中的储存能量用于通过一组涡轮机加热水，以产生高压、高温蒸汽。涡轮机组将热能转化成旋转能量，然后将废蒸汽冷凝水返回，以在锅炉内继续循环。

燃气轮机没有水加热的中间阶段，而是直接使热的高压燃烧气体通过一系列涡轮叶片。联合循环涡轮机可以通过最初使用燃气轮机，然后将废气用于加热水作为机组涡轮机进给，以提高效率。

由于将设备提升到工作温度需要时间，蒸汽轮机热发电通常需要相当长的时间以在初始启动后上升到全功率输出。在恒定输出下运行一段时间后，还需要时间来响应信号以使功率输出上升或下降。这限制了蒸汽热力发电承担额外负载的能力，因为这些额外负载出现时很少会有预警。

2. 光伏

光伏发电是通过使用半导电材料将光频谱中的能量直接转化为电能。光伏发电产生直流电力。这意味着电流在每秒 50 转内不交替方向，与传统的交流发电不一样。逆变器用来连接到电力系统。

逆变器可以提供快速的电压控制，即使在晚上没有太阳时。光伏发电的优点在于，由于消费者需求通常是由温度驱动的，因此在系统峰值负荷期间，光伏发电很可能可用。

尽管逆变器可以在没有光伏系统输出的情况下提供电压控制，但它们通常不会发生。随着更多的光伏系统连接到配电系统，因此，在国内连接的光伏系统中，故障穿越能力（在故障期间保持连接能力）存在一些问题。此外，光伏系统的输出功率是相当可变的，尤其是在阴天。因此，电力系统需要充分设计，以应对无法控制的光伏输出变化。

3. 风能

“风力发电场”通常由众多的涡轮机组成，分布在一个适度的地理区域，每台涡轮机产生少量电量，目前为 0.5～3 兆瓦单位。

由于风能是一种可再生资源，这意味着没有燃料成本，所以有初始的资本成本和持续的维护成本即可。风力发电有一个基于风速的可变输出，所以不能用常规技术来调度。当相对于整个联合发系统，风力发电的比例很高时，这是一个特别的问题。此时，必须提供其他类型的发电，并且需要在风速和发电量意外下降时介入。

传统上，风力发电场使用了感应发电机；但是，除非有显著的无功支持，否则这些将不再满足技术规则的要求。

使用双馈感应发电机（doubly-fed induction generator，DFIG）的风力发电场可以通常在有限的范围内提供电压控制，而且可能不需要额外的无功支持来满足技术规则的要求。原则上，逆变器连接的机器即使在没有风的情况下也能提供完全的电压控制，但这通常要提供额外成本。

4. 热电联产

对蒸汽或热水有需求的工商业工厂，现在通常包含了利用、生产蒸汽来改善整体经济的发电厂，例如实施“热电联产”计划的发电厂。

发电厂通常会与公用事业系统相互连接，这样的发电方式被称为“嵌入式”发电。发电厂可以输出过剩的电力，公用事业也能减少电力的进口。直接与电力供热系统耦合的热电联产方案，在满足电力系统需求的同时，也不需要将热量输出到电站其他位置。该方案可以安装一个辅助点火系统，使发电机即使在工厂的热电需求很高的时候也能有最低的电力输出。

5. 水能

水力发电利用驱动轴连接发电机组的水轮机将势能转换成电能。水轮机可以设计成用不同水位差发电。发电机组对功率输出的响应能力随水轮机组和压力机组的设计而变化。

与热力发电相比，水力发电对小负载波动的响应通常更快，并且可以更快地联机。

6.分布式发电

“分布式发电”涉及小规模发电技术与配电网的连接。小规模发电的发电厂通常位于其用户的附近。其电力通常消耗在与之连接的配电网上，而不必为了长距离传输而增加电压。

现有的配电网络已经在分布式发电占比很少的环境中发展起来。现有的配电网的设计利用了简化方式（例如，缩减配电网末端的容量，降低故障等级，使用比现有输电网上使用的更简单的控制和保护系统）。然而，增加小规模发电技术的使用并将大量的发电设备连接到现有的配电网中，这往往是有问题的。出现的安全问题包括：是否确保充分清除故障，以及与电压分布、导体载流量和故障等级相关的问题。许多现有的配电网需要改良才能满足新式发电能源的需要。

第五章补充阅读参考译文：电力质量

电力系统需要以稳定的状态运行，并且能够以可接受的质量向用户提供电力。所有连接的设备必须以所需的方式运行。

电网的正确运行要求电压保持在所有电气设备设定处理的运行范围内。概括地说，基本特性是：

- 正弦形式，具有低水平干扰（来自电动机启动、谐波等），
- 恒定的电源系统频率为 50 Hz。

在电力系统负载较重时，流过任何设备的电流不应超过其设计额定水平。在发生故障时，设备必须能够处理电流。发电机、变压器和传输电路的阻抗应该足够低，以便能够流动到所需的位置，而不会从发电到负载的压降过大。

阻抗应该足以确保任何一台设备的故障电流不会过大。新发电机有责任确保它们能够在其输出范围内输出功率，同时满足技术规则。

1.电压

电网设定的工作电压称为“标称电压”。标称电压也可以指每 1 或 100% 功率单位（power unit, pu）电压。当电网点在 1.05 或 105% pu 运行时，这意味着在该点测得的电压比标称电压高 5%。电力系统电压根据负载位置和

大小而变化，电力路径必须从发电机到负荷。

正常情况下，如果网络上任何一点的电压上升到110%以上或下降到90%以下，则交付的电力质量是不可接受的，因为它可能会损坏连接的设备。

正常运行时更接近电力用户所需，可接受的电压变化范围为正负6%标称电压。在系统负载条件下根据系统的负载情况，变压器通过改变其电压转换比（称为“分接头切换”），以使系统保持在电压范围内，在电压调节中发挥着至关重要的作用。然而，变压器需要时间来改变分接头，并且不能消除浪涌和其他短期电压偏差。

如果电压变化不规律，网络上的电源传输质量也可能很差，最常见的问题是：

• 闪烁　连接到网络上的大负载会有规律地发生变化或切换，这会导致周围网络上的电压出现小的变化。这些导致白炽灯的亮度变化，这对用户来说是恼人的问题。

• 骤降　在弱电网上启动大型电机，通常是配电网，可导致电压大幅度跌落长达十秒。电压的下降会导致附近连接的其他负载关闭。严重的情况下，可能造成损害。

• 谐波　大型变速驱动器与电网的连接将电源噪声引入到周围电网，称为谐波。谐波引起的电流会导致其他电气设备过热并发生故障。系统中长时间存在的谐波会缩短敏感设备的使用寿命。

• 不平衡　电压不平衡将导致三相电机发热并降低电机输出转矩，从而导致机器损耗增加和效率降低。

电压不平衡的一些原因是：电力系统三相中的每一相的负载不均衡；不完整的长输电线路和铁路牵引负载。

2. 频率

电力系统的频率与电力系统连接的机器转速直接相关。频率则需要维持以免损坏电机。一般来说，如果由于负荷的增加而频率下降，发电机必须增加动力，将频率恢复到正常水平，并且反之亦然。发电需要覆盖所有范围内可运行的网路频率，因为发电机计划外的错误会降低系统的频率。

3. 电流

电流通过网络中的规定回路时，不应超过相连接设备的额定功率。超过额定功率的电流会导致设备承受过多的机械应力和过热的压力。这可能会对设备造成永久性损坏。

在电路中某个点的故障会导致大电流在网络中流向故障点，该故障的设备需要能够处理大电流而不过热，直到保护能够隔离故障点的电力供应。

4.阻抗

电路的阻抗要足够低，这样发电机就能提供最大的输出功率，而不会导致电压在任何一端都过低。有时，通常在现有发电机附近，当新发电机连接到电网上的强点时，可能需要额外的阻抗以防止发电机故障危害作用超过附近电网设备的故障额定值。

5.功率

有时，需要对发电机的输出功率进行限制。这可能发生在电路故障期间，或者当系统电压达到允许范围内的极限时，将发电机的功率转移到负载中心。典型的运行约束通常是在安装新的发电机之前通过模拟一系列的操作场景来发现的。如果限制太严格，可以安装其他电力设备来增加操作范围。

6.相量功率或无功功率

“相量功率”(单位为千伏安，kVA)是在电力系统实现能量传输所需要的(无功的，或虚的部分)功率和可用电力能够从产生到传输至负载(实部)可使用功率的组合。单独讨论有功和无功功率，因为它们在电力系统中的表现不同。

6.1　*无功功率*

“无功(或者虚)功率”(单位为乏或千乏，Var 或 kVar)提供电在电力系统中传输的力。它支持将真正的电源通过电路驱动到负载区域所需的电压。变压器和许多电机吸收无功功率，并利用它建立磁场，允许能量通过磁耦合。发电机和电容器产生无功功率，其输出平衡，以满足电力系统的总无功需求。发电机可以通过改变转子的励磁来提供无功功率的可变水平，并且可以通过电压监控控制电路提供连续的电压控制。电容器是一种固定的设备，可以分区切换，提供一定的无功功率，并用于提升电压。

稳定电压需要无功支持。如果没有足够的无功功率，电压就会降低。如果差额足够大，就会导致电压崩溃。如果这里有太多的无功功率，那么就会出现过压问题。这是一个特殊的问题伴随着长而轻负荷线。

6.2　*有功功率*

“有功功率”(单位为瓦或千瓦，W 或 kW)是指电力系统可由负荷转换为其他形式能量的电力。发电的实际输出必须通过负荷和损耗消耗的功率来平衡。如果发电量大于所用能量，频率就会增加。如果发电量小于所用能量，频率就会下降。为了保持恒定的电力系统频率，发电必须与负荷和损耗完全匹配。

第六章补充阅读参考译文：输电和配电系统

在道路网络中，根据具体情况使用了一系列的设计。高速公路允许许多车辆同时行驶，有许多车道（因此即使一条车道由于事故或维修而关闭，车辆仍有可能行驶），而且不会受到交通管制的干扰。进出高速公路的较小区域道路安装了交通灯，具备较低的承载能力。在郊区，车道更窄，停车标志也很常见。电力系统也有类似的特点。一个重要的区分在于输电和配电子系统。

1. 为什么我们需要不同的电压

低电压一般用于低功耗应用，因为它们为大多数消费者需求提供足够的能量，而不会在故障期间造成过度破坏。

中等电压和相对较高的电流用于发电和负载以增强磁耦合。这允许在机械和电气形式之间转换大量电力。

在输电中使用高电压，使得大流量的电力以相对较小的电压降输送到中等距离。配电电压允许经济型配置部署在需要使用大量设备的广阔区域。并且容许相对较高水平的损耗可以节省资金成本。

使用超高电压输电能够使流过该线路的每安培电流更大。由于线路损耗与电流的平方成正比，并且不受系统电压增加的影响，所以电力可以更有效地传输，每兆瓦传输的传输线长度损失更少。这意味着在更高的系统电压下，相同导体尺寸可以传输更多的电力。这使得超高压系统可以更经济地传输大量电力。运行在特高电压下的传输线构成了大功率系统的骨干。他们能够在长距离上传输大量电力，而不会损失过多的电力。

2. 输电电路

传输电路是两个点（有时是三个）之间电能传输的连接。三相交流电力系统的一个电路将具有三个导体。

将两个点通过多个电路连接在一起是有利的，因为如果一个电路出现故障或因维护而停止工作，这可以避免供电中断。

输电电路可以是架空的，也可以是地下的，也可以是两者的组合。

二次输电系统将电力从负载中心（或“供应点”）配送给具体的局部地区

变电站(也称为区域变电站),在那里电力被进一步分配。

诸如工业场所或大型商业场所的大型电力用户可能需要他们自己的区域变电站,在这种情况下,他们有一个专用的子传输连接到他们的处所。

配电系统将电力从区域变电站分配到路边变电站或其他本地网络,形成最终连接,从而将电力降至可安全使用的电压等级。大多数情况下,这些系统是每个区域变电站的辐射形式,在相邻网络的馈线之间有一个开放点,允许在相邻的供电点之间转移负载。

3. 电力线路

3.1　架空线

为了避免短路或故障,架空输电线路的电塔采用铁制或木制,电线杆采用混凝土或金属,从而将导体保持在地面以上的安全距离。一堆绝缘体处在电导体与安装在电线杆或电塔上的横臂之间。这些绝缘体能够防止电流传输到电线杆或者地面上,从而避免产生故障。

3.2　电缆

电缆包围着电导体,每个电导体都配有绝缘材料和接地导电屏蔽体。这样的搭配使得电缆可以安装在建筑里或直接在地面上,也不会发生导体短路。

3.3　高压直流输电

HVDC 意为高压直流输电。高压直流输电分为“传统高压直流输电”和“柔性高压直流输电”两种。对于远程传输从几百兆瓦到近三千兆瓦的大规模电量,传统高压直流输电很有效。成套设备需要安装高压直流输电电路,并且不存在与短距离交流输电线路的经济竞争。为了远距离(至少数百公里)传输高压电,高压直流输电提供了以下有利条件:

- 只需要两个电阻(如果地面作为回路甚至只需要一个电阻)。
- 更好的系统稳定性,即两条既不同步也不同频的电网能够被连接。

第七章补充阅读参考译文:变电站设备——变压器/断路器/母线/隔离开关

变电站是电力系统上的连接点,她能够连接多重电路,也可以将电压转换为更可用的等级。变电站的一般用途为:

• 通过升压变压器将发电机连接到输电网；

• 通过降压变压器连接客户网络；

• 能够在多个电网中共享电力；

• 如果已连接的电路发生了故障，变电站能够清理故障并且重新分配电力，从而最小化电力供应中断的影响；

• 在不中断电力供应的情况下有计划地断开电路，从而维修电力系统设施。

一个变电站可能是完全由安装在电线杆上的一个变压器和两个转换器组成的。然而，一个特定电网的主变电站更可能包括以下三个设备的场所，分别是控制室、室内、室外或者同时控制室内外的接电装置以及变压器。

1. 变压器

变压器可以提高网络的电压水平，以便远距离传输大量电力，然后将电压降低到消费者可以使用的水平。

变压器有两个用途：它们可以跨越两个不同电压系统传输电力，和/或它们可以提供系统之间的电力隔离。

双绕组变压器可将一个电压等级的网络连接至初级绕组，并将另一个网络连接至次级绕组。变压器通常在同一个外壳中包含所有的三相，因此有三个初级绕组(每相一个)和三个次级绕组。

初级绕组和次级绕组上的匝数之比决定了电压转换比。

由于传输电压可能因网络负载而异，因此可以使用分接头切换器对变压器的电压转换比率进行微调。分接头切换器可以增加或减少变压器一侧的匝数，以改变电压转换比。其意义在于，当变压器调节次级网络的电压时，主网络的电压可能会发生变化。初级和次级网络之间的电隔离通过在初级和次级侧上的导体之间没有直接电连接来实现。此外，磁耦合阻止了电流流动量的变化(呈现为阻抗)，这限制了故障电流的大小。有三种主要类型的变压器用于输电网之间的电力传输。

• 发电机升压变压器或发电机变压器　发电机变压器用于接受中等电压下产生的大电流，并将电力以高电压和相对低的电流传输到传输系统。发电机变压器通常具有较大的电压比，并在发电机侧具有三角形连接，以限制零序电流的流动，从而限制发电机接地故障。

• 互连变压器　互连变压器在不同电压的传输网络之间传输功率。这些变压器的电压比较小。

• 降压变压器　降压变压器将电压从一个水平降低到另一个水平。降压变压器有两大类：

• 供电（区域变电站）变压器　将电压从传输或子传输高压降低到配电电压水平。

• 配电变压器　将电压从配电高压降低到低压。

电力系统中使用的其他类型的变压器包括接地变压器，电流互感器和电压互感器。

2.断路器/母线/隔离开关

断路器（通常缩写为“CB”）通常是能够在变电站中切断故障电流的开关。在发生故障时，断路器可以非常迅速地切换，以最大限度地减少故障位置和所有连接设备发生的损坏量，并最大限度地减少对客户的干扰。

在故障情况下，可能会有大电流（可能由于短路）流过断路器。为了成功地清除故障，必须将断路器设计成能够在最大电流流动期间打开开关（或断开连接）。

母线描述了将变电站内所有电路连接在一起的电导体。它允许来自任何连接电路的电流流向任何其他电路。不能直接控制通过母线的电流，但设备的切换可以改变电流。不同的母线配置用于不同的环境。

隔离开关在电气网络上的两点之间提供了必要的、可见的电隔离间隙。

如果需要在输电线路上进行维护，从理论上讲，为了安全起见，维护人员的每一端都要打开断路器并将其连接到地线就足够了。然而，与隔离开关不同，断路器不会在网络中出现可见的中断，并且不足以向现场人员证明网络部分是隔离的。

隔离开关还提供了隔离断路器两个端子的机构，从而可以对断路器本身进行维护。隔离开关还可以允许在故障发生后快速恢复所有非故障设备。这在处理断路器故障时尤为重要。

第八章补充阅读参考译文：变电站设备：接地开关，电流互感器/电压互感器

1.接地开关

接地开关使电路的一部分进入土壤，成为一种在维护过程中的一种安全

性预防手段。接地开关会保护人们在维护一条电路时不受因疏忽大意导致的突然通电带来的瞬间危害;并且会为所有那些在邻近其他电路流动的杂散电流提供一条通路。

2. 电流互感器/电压互感器

互感器塑造了能源系统的眼睛和耳朵。这些装置保证了对主要导体的电流和电压的安全测量。顾名思义,这些测量装置将他们要测量的电学性质向下变换一个层级,以适应输入电子装置进行分析,同时与电力系统绝缘的测量装置。这些装置将信息传递给监控系统的继电器;同时还传递给操作员,由他们来管理电功率通量。

电流互感器往往使用初级导体作为一次绕组。它通过一个放在主导体周围的环形铁心所缠绕而成的次级线圈来测量电流。次级线圈匝数决定电流的减少比率;而选取该比率将用于减少正常运行电流直到安全装置可以进行测量的程度。

一个开路的电流互感器的次级绕组非常危险,因为次级绕组成了电流源。这意味着次级绕组会在任何物质形式的开放电路中使电流通过。它可以向临近的人体传递致命的电流,并且可以毁坏次级线圈的绝缘体。

一个电压互感器与降压变压器的运行类似,因为它会将任何系统电压(如 6.6~330 kV)下降到 110 V,然而,二者的主要区别是电力变压器不是用于能源转化,而仅仅是用于对准确电压测量有必要的电能最小值进行转化。

第九章补充阅读参考译文:变电站设备——电涌放电器、无功功率支持、接地

1. 电涌放电器

电涌放电器实际上是一种对于电力系统的电压释放装置。它们在电压变得过高时将电流通过自己流向地面。电涌放电器在所有种类的伤害中都适用。电涌放电器与短路并不相同,当电流穿过电涌放电器流向地面时它并没有将连接电力网络的电压等级下降到低于它的额定运行值。

2. 无功功率支持

无功功率为电力系统提供所需的电压支持,以将实际功率从发电输送到

负载。

无功功率支持可以安装在网络的任何位置(尽管变电站通常是被选定的)来维持周围的网络稳定性并确保其正常运行。常见的无功功率支持包括:

• 电容器组。在距离发电较远的节点上的电力系统负载较大时,电压可能会下降到不佳的运行水平。在极端情况下,它可能导致预料之外的负载跳闸。

电压升高是安装电容器的净效应,这有助于缓解在重载时的电压下降现象。而电压降低是由于在供电电路中无功功率的增加,以及沿着线路上的压降随之增加后引起的。如果无功功率的流向反转,电容器同样也会导致电压升高。

• 并联电抗器。当处于轻负载期或是当网络中的供电节点随着大规模电量的流出而电压上升时,一个并联电抗器可以防止在部分网络中发生的过电压事故。

• 静态无功补偿器(SVC)。SVC 允许提供或消耗可变数量的无功功率,因此它被用于控制网路上某节点的电压,使之能够在一系列不同的运行情况下达到期望的水平。

3. 接地

在大多数变电站中,所有带有金属部件的设备都不属于主电路部件,它们都接地以确保安全,并为电力系统提供一个通用参考。

在大型变电站中,与地面的物理连接通常由埋铜的导体像网格一样连接在一起。在更小的设备中或杆塔中,接地连接可以为一根插入地下的单针。

第十章补充阅读参考译文:变电站设备——继电器室和变电站接线配置

1. 继电器室

继电器室包含变电站的“大脑”。继电器室有三个关键功能,来确保变电站可靠和安全地运行:

• 控制。变电站的控制功能允许运营商在变电站内运行设备。如果需要在某一台设备上进行维护,操作人员可以切换变电站的各个部分,以确保设备的安全工作，同时保持对用户的供电。

如果连接到变电站的电路发生故障，保护系统将自动断开故障设备。然后工作人员可以重新配置网络以恢复供应。

• 保护。保护的作用是启动断路器的运行尽可能快地在变电站或其附近清除故障，以尽量减少对设备的损坏和对用户断电的影响。

• 通信。变电站安装有通信设备，因此可以从中央控制室观察到多个变电站的电力流状态。这样就可以对整个电力系统进行监控；通信对发电调度和电力系统充足储备容量的分配起着至关重要的作用。监控和数据采集系统（SCADA）接口构成电力系统通信系统的主要支柱。通信也用于帮助保护设备的运行。通信是用于通信或广播信息的系统和过程，特别是通过电力或无线电波的手段。

由于控制室的电源对于管理通过变电站的电力流动至关重要，因此在常见的停电期间有备用电池电源可以维持足够的运行时间。

2. 变电站接线配置

不同的变电站接线配置用于不同重要级别的变电站。对于 330 kV 母线，有些电力公司至少需要一个网格接线配置来提供一定的安全性以防止发生总线故障，来满足技术规范对可靠性的要求。网格接线配置不容易扩展，因此如果网站有六个以上的交换机，则使用一台断路器和半接线配置。

2.1　单母线配置（每个电路一台断路器）

单母线的布置操作简单，对信号的依赖性最小，能合理的保护操作，并且有利于经济地增加未来的馈线托架。

单母线布置通常在每个连接的电路上都会有一个断路器，这将在外部故障期间保护变电站免于中断。但是，总线故障将导致变电站完全中断，直到故障原因得以解决。

通过安装“母线分段断路器”，单母线布置可以提供额外的安全供电。母线分段断路器将母线分成两部分，在任一部分发生母线故障后，另一部分母线继续工作。

2.2　网格接线配置

一个接线配置为网格的变电站具有与电路数量相同的断路器数量。该方案比没有母线断路器的单母线提供更好的功能和设施。但是，除非最初的设计适合未来的扩展，否则在需要附加电路的情况下不能轻易扩展。网格变电站通常限于六个电路。

2.3　断路器和半接线配置(每个电路一台半断路器)

“断路器和半接线配置”使用三台断路器连接两个电路(或每个电路一台半断路器)。这种接线布置提供了双母线布置的好处,并且提供了更高的安全等级。断路器和半接线配置具有每个电路在母线之间切换而不停电的能力。单一母线断电故障时,供电继续维持。取决于断路器的布置,两个母线的断电故障时,一定水平的电力供应也能够维持。

这种接线配置提供与双母线双断路器接线配置相同的可靠性水平,并且断路器更少。

第十一章补充阅读参考译文:电力系统故障

电力系统故障是由于主电导体(或多个导体)之间的绝缘击穿导致正常电力流动的中断而引起的。这些事件是不可预测和避免的。为了避免损坏网络的其余部分,并且(进一步)避免中断供电,故障组件被“保护系统”分离。

可能导致绝缘破坏的因素包括:

- 大气条件,如雷电、风等
- 暴露绝缘体上的污染物形成电流泄漏路径
- 虫类
- 设备故障
- 人为错误(错误切换)

1.故障类型

有两种类型的故障:“平衡故障”和“不平衡故障”。

• 平衡故障。三相故障涉及一个电路的所有三相的短路。三相故障也被称为平衡故障,因为所有故障电流都流经导体而不流过地面;并且所有三个相均受故障均匀影响。系统上的三相故障通常会给电力系统造成很大程度的中断。这是因为短路导致电压和相应的功率传输能力在所有三相上崩溃。

• 不平衡故障。不平衡故障最为常见,通常由“单相接地”故障引起。不平衡故障是指导致故障电流在故障电路的三相中不均匀流动的任何故障。最常见的一种不平衡故障是“单相接地”故障。在传输线上尤其如此,在传输线上两相间短路的可能性较低,这是由于各相之间的间隔很大。

2.故障级别

终端变电站内的单相接地故障比三相故障具有更高的故障电流;然而单相接地故障通常会对整体功率传输能力造成较小的中断,因为有两个"健康"相能够传输有限的功率。

•故障前。发生故障时,电力系统的"故障前"情况很大程度上决定了电流的大小。

•次瞬态。一旦故障已经建立,"次瞬态阶段"处理流入故障的电流的初始电涌。此阶段内的当前峰值决定了设备上的机械冲击负荷。当前峰值中包含的能量决定了设备过热的速度。

•瞬态。"瞬态阶段"描述故障发生之后发生的电流峰值的尾端。断路器通常需要在此阶段断开电流以隔离故障并帮助系统恢复。然而,断路器和继电器的技术进步使得响应更加灵敏,并且可能要求断路器在亚瞬态电平下达到分断电流。

•稳态。"稳态故障电流"表示在初始电涌稳定后,如果故障在系统上保持未清除状态还将流动的电流。在传输电路上很少观察到故障的稳定状态,因为在初始电涌期间通常会清除故障。但是,在配电线路中,系统上的远程故障更可能出现类似于大负载的情况。故障可能需要较长时间才能消除,因此接近稳态电流值。

3.影响故障电流的因素

在故障点上流动的总电流是由电网许多电源的贡献组成的。产生的电流并不仅仅来自发电机;由于储存在电机的旋转磁场中的能量,电动机也可能产生到故障点的电流。

一般来说,发生在发电位置附近的故障会有高的故障电流,而在距离较远的电网点上的故障会有较低的故障电流。

•阻抗。在故障点和电网上发电点之间的线路阻抗,很大程度上决定了故障电流的值。从故障到供电点的阻抗称为"源阻抗"(有时称为系统阻抗)。

如果特定电网点的系统阻抗很小,那么当故障发生时,该点将具有较大的故障电流。阻抗的大小取决于该点连接的电路数量和电压电平。具有小功率传输容量的点将具有高系统阻抗。初始峰值故障电流也受到源阻抗的电阻与电抗之间的比值影响。如果系统电抗明显大于系统电阻,那么初始直流电流将持续较长时间。

•电网强度。电网点的故障级别决定了点的强度。如果一个电网点有一个故障，就会有一个大的故障电流，据说它有一个“高故障级别”，而相反，一个带有小故障电流的电网点可以被称为带有“低故障级”别的点。电网点的故障级别决定了该点的强度，或者是指在稳定条件下传输功率的能力。

4.故障等级

可以从稳定性角度看出，一个新发电机连接到电网上的一个强点（高故障等级），输出功率的问题要比在电网上连接弱点（低故障等级）的要小。

为了限制故障的范围，所有非故障的设备都需要能够处理流向故障位置的电流，直到故障可以安全清除。如果不是这样，一个错误就会导致级联故障和对电网周围环境的破坏。

将新发电机添加到现有电网可能会导致故障级别增加，超出目前连接的设备的故障等级。在这种情况下，可能需要换用更高的故障等级的设备，或使用设备减少来自发电机的故障电流影响。

第十二章补充阅读参考译文：接地

1.什么是接地？

几乎只要有大型电力系统，地面就会被用作测量电力系统电压的参考点。一般来说，地面在任何地理区域都具有相同的电压，因为它是非常好的导体（除了某些地区有非金属岩层）。连接到电力系统的设备接地（通过金属导体连接到地），以确保设备安全可靠地运行。

2.为什么要求接地？

电力系统设备的接地很重要，主要有两个原因：

• 当发生绝缘故障时，一段电流将安全地转入地下，不会对人员或其他设备造成损害。

• 与地面接触的带电导体故障可以被更容易地检测到。易于通过保护，更快速地清除故障并保护电力系统中的其他设备。

3.接地

在一个变电站安装一个接地栅极，以限制在大量电流通过地面本身流入或流出接地栅极时，周围区域地面电压的升高。

接地栅极还消除了变电站周围的电压上升，使得没有电流流过变电站的一般区域中的工作人员。

4.变压器和发电机接地

电力系统变压器的中性点接地,以便在故障时允许不平衡的接地电流流动,由此在稳定阶段限制电压上升,允许安装选择接地故障继电器。供电变压器、发电机变压器和配电变压器通常(但不是全部)具有星形连接的次级绕组,具有中性点直接接地和三角形连接的基极。

接地方法决定了系统在接地故障时的表现。

• “直接”或“有效接地”会导致高接地故障电流,并在故障出现时导致电压急剧下降。中性点基准相对于无故障相位大致保持固定,因此在未故障相位上接地电压几乎没有上升。

• “高阻抗系统接地”将导致最小的接地电流。随着中性点参考电位与故障相电位相等,无故障相的相对地电压可以上升至故障前相间电压。在扩展系统中,系统电抗和电容可以结合起来,进一步提高电压,有时达到危险水平。

• “低阻抗接地”越来越多地被使用。由于接地阻抗较小,故障相的相对地电压将上升,但不会上升至故障前的相间电压值。

发电机已经经历了定子接地技术的演变。具有自己的机组变压器的大型发电机现在几乎普遍具有高电阻接地,通过将故障电流限制在非常低的水平来限制定子损坏。直接连接到 11 kV 配电系统的小型发电机通常是低电阻接地。

变压器结构矢量组对确定系统接地的有效性具有重要意义。具有三角形/星形结构的变压器,在星形侧提供稳固的接地点,但在三角形侧没有。具有一个接地中性点的星形/星形变压器可以在某些情况下提供足够的接地,例如三相变压器的中性点通量可以通过变压器油箱,即该油箱用作三角形绕组。

第十三章补充阅读参考译文:防雷装置和电涌保护器

1.防雷装置

由于大多数电力系统都具有架空电线和承载这些电线的高金属结构的网络,所以该系统容易受到雷击。雷击电路中的任何导体会引起电功率通量的激增，并可能导致电路沿线设备的级联故障。

防雷保护是将雷击直接接地，不通过主电路，以此保护电力设备。

防雷系统由三部分组成：

- 空气终端系统，由放置在电力导体上方的电线组成，以捕捉雷击。
- 接地引下线，即将电流从架空电线接地的钢塔和/或导体。
- 接地系统，使电流从下导体流入地面。接地系统可以是接地栅极，混凝土中使用的强化钢筋、接地销头等。

2. 电涌保护器

由于故障变压器需要很长的维修时间和大量的维修费用，电涌保护器通常安装在变压器附近，以防止变压器的电涌。为了防止变压器产生绝缘故障，电涌保护装置一般安装在变压器附近。它们也安装在变电站进线的线路上，以防止损坏仪表变压器。电涌放电器是电力系统压力释放的一种形式。如果电压变得太高，那么允许电流通过它们流入大地。电涌放电器非常适用于应对各种损坏。

由于雷击（如果没有得到防雷保护的充分保护）或开关电涌，电压和电流可能会沿电路流动，所以电涌放电器可以安装在电力系统上，以保护设备不受峰值电压和电流的冲击。

参考文献

[1] Ackermann, T. (Ed.). (2005). *Wind power in power systems* (Vol. 140). Chichester: John Wiley.

[2] Billings, K. H., & Morey, T. (2011). *Switchmode power supply handbook*. McGraw-Hill.

[3] Bird, L., Milligan, M., & Lew, D. (2013). *Integrating variable renewable energy: Challenges and solutions* (No. NREL/TP-6A20-60451). National Renewable Energy Lab. (NREL), Golden, CO (United States).

[4] Catford, J. C. (1965). *A linguistic theory of translation: an essay in applied linguistics*. Oxford University Press.

[5] Chattopadhyay, P. (2013). *Boiler operation engineering: questions and answers*. McGraw-Hill Education.

[6] Emadi, A., Nasiri, A., & Bekiarov, S. B. (2017). *Uninterruptible power supplies and active filters*. CRC press.

[7] Gieras, J. F. (2009). *Permanent magnet motor technology: design and applications*. CRC press.

[8] Goutsos, D., & Georgakopoulou, A. (1997). *Discourse analysis: an introduction*. Edinburgh University Press.

[9] Von Meier, A. (2006). *Electric power systems: a conceptual introduction*. John Wiley & Sons.

[10] Keller, K. (2010). *Electrical Safety Code Manual: a plain language guide to National Electrical Code, OSHA and NFPA 70E*. Butterworth-Heinemann.

[11] Klass, D. L. (1998). *Biomass for renewable energy, fuels, and chemicals*. Elsevier.

[12] Kusko, A., & Thompson, M. T. (2007). *Power quality in electrical systems* (Vol. 23). New York: McGraw-Hill.

[13] Mallick, A. R. (2015). *Practical boiler operation engineering and power plant*. PHI Learning Pvt. Ltd..

[14] Mann, F. M. (1998). *Hanford immobilized low-activity tank waste performance assessment* (No. DOE/RL-97-69). Fluor Daniel Hanford Inc., Richland, WA (United States).

[15] Newmark, P. (1981). *Approaches to translation*. Oxford: Pergamon Press.

[16] Newmark, P. (1991). *About translation* (Vol. 74). Multilingual matters.

[17] Nida, E. A., & Taber, C. R. (2003). *The theory and practice of translation* (Vol. 8). Brill.

[18] Ockwell, D., & Byrne, R. (2016). *Sustainable energy for all: innovation, technology and pro-poor green transformations*. Routledge.

[19] Pym, A. (Ed.). (2016). *The return to ethics: special issue of The Translator* (Vol. 2001). Routledge.

[20] Rumelhart, D. E. (1980). On evaluating story grammars. *Cognitive Science*, 4(3), 313-316.

[21] Vermeer, H. J. (1978). *A framework for a general theory of translation*. Heidelberg: Heidelberg University, 67.

[22] Yoshio, M., Brodd, R. J., & Kozawa, A. (2009). *Lithium-ion batteries* (Vol. 1). New York: Springer.

[23] 方梦之.(2005).英汉翻译基础教程.北京:中国对外翻译出版公司.

[24] 格特.(2004).翻译与关联:认知与语境.上海:上海外语教育出版社.

[25] 龚明星.(2011).英汉翻译中的增词与减词探究.吉林广播电视大学学报,(6),136-137.

[26] 郭着章,李庆生.(2003).英汉互译实用教程.武汉:武汉大学出版社.

[27] 胡庚申.(2013).生态翻译学:建构与诠释.北京:商务印书馆.

[28] 胡壮麟.(1994).语篇的衔接与连贯.上海:上海外语教育出版社.

[29] 黄忠廉.(2002).变译理论.北京:中国对外翻译出版公司.

[30] 连淑能.(2010).英汉对比研究(增订本)(Vol. 9).北京:高等教育

出版社.
[31] 沈苏儒.(1998).论信达雅：严复翻译理论研究.上海:商务印书馆.
[32] 夏廷德.(2006).翻译补偿研究.武汉:湖北教育出版社.
[33] 杨丰宁.(2006).英汉语言比较与翻译.天津:天津大学出版社.
[34] 于建平.(2002).科技语篇的主位—述位推进模式.中国科技翻译,(2),3-6.
[35] 余承法.(2013).全译求化机制论(博士论文,华中师范大学).